Computer Simulation in Chemical Physics

Computer Simulation in Chemical Physics

Contributors
_____

Alexander Pertsin, Dmitry Platonov et al.

AURIS
Reference

www.aurisreference.com

Computer Simulation in Chemical Physics

Contributors: Alexander Pertsin, Dmitry Platonov et al.

Published by Auris Reference Limited

www.aurisreference.com

United Kingdom

Computer Simulation in Chemical Physics

ISBN: 978-1-78154-901-8

British Library Cataloguing in Publication Data
A CIP record for this book is available from the British Library

Printed in the United Kingdom

Exclusively distributed by CBS Publishers & Distributors Pvt. Ltd.

Sales & Distribution Rights only for India, Pakistan, Bangladesh, Sri Lanka, Nepal and Bhutan.This book is not to be sold outside these territories.

Contents

List of Abbreviations

CDM	Cis-dimer model
DAH	Differential Adhesion Hypothesis
DHFR	Dihydrofolate reductase
DDSCAT	Discrete dipole scattering
DPD	Dissipative Particle Dynamics
DG	Distance geometry
EELS	Electron energy loss spectroscopy
EDTA	Ethylenediaminetetraacetic acid
FEM	Finite element method
GGH	Glazier-Graner-Hogeweg
HDPE	High density polyethylene
KMC	Kinetic Monte Carlo
MD	Molecular dynamics
MC	Monte Carlo
NPMs	Nanoporous metals
NTA	Nitrilotriacetic acid
NWHBL	Normalized weighted heterotypic boundary length
NOE	Nuclear Overhauser
rms	Root mean square
SM	Saturation model
STM	Scanning tunneling microscopy
SPC	Simple three-point charge
SD	Stochastic Dynamics
SPR	Surface plasmon resonance
SERS	Surface-enhanced Raman scattering
D	Three-dimensional
TEM	Transmission electron microscopy
DNMR	Twodimensional nuclear magnetic resonance spectroscopy
WHBL	Weighted heterotypic boundary length ,

List of Contributors

Alexander Pertsin
Angewandte Physikalische Chemie, Universität Heidelberg, Im Neuenheimer
Feld 253, 69120 Heidelberg, Germany

Dmitry Platonov
Angewandte Physikalische Chemie, Universität Heidelberg, Im Neuenheimer
Feld 253, 69120 Heidelberg, Germany

Michael Grunze
Angewandte Physikalische Chemie, Universität Heidelberg, Im Neuenheimer
Feld 253, 69120 Heidelberg, Germany

Michael R. Twiss
Department of Chemistry, Biology and Chemical Engineering, Ryerson Polytechnic University, 350 Victoria Street, Toronto, Ontario M5B 2K3, Canada

Olivier Errécalde
Les Laboratoires Aeterna Inc., 456 rue Marconi, Ste-Foy, Québec G1N 4A8,
Canada

Claude Fortin
Département de Protection de l'Environnement, Institut de Protection et de
Sureté nucléaire, BP No 1, 13108 Saint-Paul-lez-Durance, Cedex, France

Peter G. C. Campbell
INRS-Eau, Université du Québec, C.P. 7500, Ste-Foy, Québec G1V 4C7, Canada

Catherine Jumarie
Département des Sciences biologiques, Université du Québec à Montréal, C.P.
8888, Succursale Centre-ville, Montréal, Québec H3C 3P8, Canada

Francine Denizeau
Département de chimie, Université du Québec à Montréal, C.P. 8888, Succursale Centre-ville, Montréal, Québec H3C 3P8, Canada

Edward Berkelaar
Department of Land Resource Science, University of Guelph, Guelph, Ontario
N1G 2W1, Canada

Beverley Hale
Department of Land Resource Science, University of Guelph, Guelph, Ontario N1G 2W1, Canada

Ken van Rees
Department of Soil Science, University of Saskatchewan, Saskatoon, Saskatchewan S7N 5B3, Canada

P. A. Netz
Departamento de Química, ULBRA, Canoas RS, Brasil and Departamento de Química, Unilasalle, Canoas RS, Brasil

F. Starr
Center for Theoretical and Computational Materials Science and Polymers Division, National Institute of Standards and Technology, Gaithersburg, Maryland 20899, USA

M. C. Barbosa
Departamento de Física, UFRGS, Porto Alegre RS, Brasil

H. Eugene Stanley
Center of Polymer Studies - Boston University Boston, MA 02215, USA

Ying Zhang
Cancer and Developmental Biology Laboratory, National Cancer Institute, Frederick, Maryland, United States of America

Gilberto L. Thomas
Instituto de Fı́sica, Universidade Federal do Rio Grande do Sul, Porto Alegre, Brazil

Maciej Swat
Biocomplexity Institute and Department of Physics, Indiana University, Bloomington, Indiana, United States of America

Abbas Shirinifard
Biocomplexity Institute and Department of Physics, Indiana University, Bloomington, Indiana, United States of America

James A. Glazier
Biocomplexity Institute and Department of Physics, Indiana University, Bloomington, Indiana, United States of America

Yong-Lei Wang
Department of Materials and Environmental Chemistry, Arrhenius Laboratory, Stockholm University, Stockholm, Sweden

Fredrik Hedman
Noruna AB, Stockholm, Sweden

Massimiliano Porcu
Department of Materials and Environmental Chemistry, Arrhenius Laboratory,
Stockholm University, Stockholm, Sweden
Dipartimento di Scienze Chimiche e Geologiche, Cagliari University, Cittadel-
la Universitaria di Monserrato (CA), Monserrato, Italy

Francesca Mocci
Department of Materials and Environmental Chemistry, Arrhenius Laboratory,
Stockholm University, Stockholm, Sweden
Dipartimento di Scienze Chimiche e Geologiche, Cagliari University, Cittadel-
la Universitaria di Monserrato (CA), Monserrato, Italy

Aatto Laaksonen
Department of Materials and Environmental Chemistry, Arrhenius Laboratory,
Stockholm University, Stockholm, Sweden

Arkady M. Ilyin
Kazakh National University, Physical Department, Kazakhstan

Wilfred E van Gunsteren
Department of Physical Chemistry University of Groningen Nijenborgh 16,
NL-9747 AG Groningen (The Netherlands)

Herman J. C. Berendsen
Department of Physical Chemistry University of Groningen Nijenborgh 16,
NL-9747 AG Groningen (The Netherlands

Re Xia
Key Laboratory of Hubei Province for Water Jet Theory and New Technology,
Wuhan University, Wuhan 430072, China

Run Ni Wu
Key Laboratory of Hubei Province for Water Jet Theory and New Technology,
Wuhan University, Wuhan 430072, China

Yi Lun Liu

State Key Laboratory for Strength and Vibration of Mechanical Structures, School of Aerospace Engineering, Xi'an Jiaotong University, Xi'an 710049, China

Xiao Yu Sun

Department of Engineering Mechanics, School of Civil Engineering, Wuhan University, Wuhan 430072, China

Preface

Computer simulation is an essential tool in studying the chemistry and physics of chemicals. Simulations allow us to develop models and to test them against experimental data. These can be used to evaluate approximate theories of chemicals, and to provide detailed information on the structure and dynamics of chemical. The text *Computer Simulation in Chemical Physics* deals with computer simulations, which vary from computer programs to network-based groups of computers, in chemical physics. First chapter focuses on computer simulation of short-range repulsion between supported phospholipid membranes. In second chapter, we describe a detailed approach to coupling the techniques of toxicity testing in defined chemical media with the accurate calculation of chemical speciation using a computerized chemical speciation program. Third chapter discusses how the anomalous diffusivity is related to the structural anomalies. In fourth chapter, we present three models relating the surface density of cadherins to the net intercellular adhesion and interfacial tension for both discrete and continuous levels of cadherin expression. In fifth chapter, we show how nonequilibrium methods can be used to calculate equilibrium free energies. Non-uniform FFT and its applications in particle simulations have been presented in sixth chapter. In seventh chapter, we focus on study of possible production composite materials based on Be and Al matrices, with using graphene fragments as reinforcement elements. In last chapter, we consider the computer simulation of systems in equilibrium.

Chapter 1

COMPUTER SIMULATION OF SHORT-RANGE REPULSION BETWEEN SUPPORTED PHOSPHOLIPID MEMBRANES

Alexander Pertsin, Dmitry Platonov, and Michael Grunze

Angewandte Physikalische Chemie, Universität Heidelberg, Im Neuenheimer Feld 253, 69120 Heidelberg, Germany

ABSTRACT

The grand canonical Monte Carlo technique is used to calculate the water-mediated pressure between two supported 1,2-dilauroyl-DL-phosphatidylethanolamine DLPE membranes in the short separation range. The intra- and intermolecular interactions in the system are described with a combination of a united-atom AMBER-based force field for DLPE and a TIP4P model for water. The total pressure is analyzed in terms of its hydration component and the component due to the direct interaction between the membranes. The latter is, in addition, partitioned into the electrostatic, dispersion, and steric repulsion contributions to give an idea of their relative significance in the water-mediated intermembrane interaction. It is found that the force field used exaggerates the water affinity of the membranes, resulting in an overestimated hydration level and intermembrane pressure. The simulations of the hydrated membranes with damped water-lipid interaction potentials show that both the hydration and pressure are extremely sensitive to the strength of the water-lipid interactions. Moreover, the damping of the mixed interactions by only 10%–20% changes significantly the relative contribution of the individual pressure components to the intermembrane repulsion.

INTRODUCTION

The short-range repulsive force occurring between phospholipid membranes in water and aqueous solutions has long been of interest to physicists and biologists, mainly in the context of cell recognition, adhesion, and fusion.[1] Despite considerable progress in direct experimental measurements of intermembrane forces using the surface force apparatus SFA and atomic force microscopy, the nature of the shortrange repulsion is still poorly understood.[2] In a general case, the instantaneous force operating between two parallel membranes, m and m, across the water gap can be represented as the sum of the force exerted on membrane m by water molecules, $f_w \to m$, and the force due to direct interaction between the membranes fm'→m. Such a representation is correct inasmuch as the three- and higher-body effects in the intermolecular interaction energy can be neglected. The ensemble average of the water-membrane force, $\langle f_{w \to m} \rangle$, defines the hydration solvation pressure, $p_d = A^{-1} \langle f_{m' \to m} \rangle$, where A is the membrane area and pb is the bulk water pressure. Adding the pressure component arising from the direct intermembrane interaction, $p_d = A^{-1} \langle f_{m' \to m} \rangle$, , one gets the net pressure, $p = p_h + p_d$, as measured in a SFA experiment. To avoid misunderstanding, it is worth noting that the term "hydration pressure" or, more generally, "solvation pressure" is frequently used to mean the net pressure p and not ph. Here we prefer to follow the terminology of Evans and Marconi,3 wherein p_h refers to the genuine solvation pressure, free from the direct interaction between the surfaces that confine the solvent.

The origin of short-range intermembrane repulsion was initially associated with the hydration pressure p_h. [4-6] Considering the high hydrophilicity of the membrane surface, the repulsion was ascribed to the energy loss due to the exclusion of water and the associated dehydration of the membranes when they approached each other. The hydration repulsion had also a "structural" interpretation, which involved orientational polarization or "structuring" of water next to the membrane surface. The weaknesses involved in the hydration mechanism of the short-range repulsion were discussed at length by Israelachvili and Wennerström.2,7 It was argued, in particular, that the range of orientational polarization in water was limited by one or two water diameters, which was noticeably less than the observed range of the hydration force. This argument is particularly true of phospholipid membranes, where the range of orientational polarization is restricted by a high molecular-level roughness of the membrane surface.

The alternative explanation suggested by Israelachvili and Wennerström2,7 for the short-range repulsion implies the dominant role of direct intermembrane interactions, as characterized by p_d. In this explanation, the repulsion arises from entropy-driven deviations of hydrated membranes from ideal planar geometry. Most important deviations are thermal undulations of the membrane surface, fluctuations in the membrane thickness like peristalsis, and protrusion of individual lipid molecules and/or their headgroups into the aqueous phase. The deviations of this kind are taken into account by adding appropriate correction terms to the force calculated on the assumption of the ideal membrane geometry. Following Israelachvili and Wennerström,2,7 these corrections terms are usually referred to as the undulation, peristaltic, and protrusion forces or, together, the entropic forces. For mem- branes supported on flat rigid substrates as studied in SFA experiments, the only significant force of this kind is the protrusion force. Note that the protrusion force is not a new physical type of force. It enters into p_d provided that the ensemble averaging properly samples protruded configurations. The weaknesses of the Israelachvili and Wennerström's2,7 ideas have been discussed by Parsegian and Rand.8 It was in particular noted that the suggested dependence of the protrusion energy on the molecular out-of-plane displacement would lead to impossibly high solubilities of phospholipids in water. It was also argued that the protrusion force could not explain the observed strong effect of a single methylation of an ethanolamine group on the intermembrane force: Would the repulsion be due to molecular protrusions, it would not be so sensitive to the methylation of the headgroup.

Unfortunately, the theoretical treatments of both the hydration and entropic forces involve serious simplifications and a number of unknown parameters. The improvement and verification of theory are only possible on the basis of a molecular-level knowledge of the membrane structure along with the simultaneous knowledge of the water-mediated intermembrane force. This attaches significance to direct computer simulations of the force as the respective ensemble average. An advantage of computer simulation is the possibility to partition the calculated net force into physically distinct components associated with the individual components of the potential energy. Furthermore, the direct intermembrane pressure p_d can be decomposed into contributions from individual lipid molecules, which allows, in particular, an analysis of molecular protrusions and their contribution to the short-range repulsion. An attempt at direct computer simulation of the operative force between phospholipid membranes in water has been undertaken in our recent work9 using the grand canonical Monte Carlo GCMC technique. The configuration of the model system was similar to that of a SFA experiment Fig. 1: Two 1,2-dilauroyl-DL-phosphatidylethanolamine DLPE membranes were supported on parallel solid substrates and brought to equilibrium with bulk water at ambient conditions. The starting membrane structure was constructed based on the x-ray diffraction data for the DLPE-acetic acid crystal.10 During the simulations, the area per DLPE molecule was kept fixed at the respective crystal-state value, A= 38.6 Å2 . The net intermembrane pressure p was analyzed in terms of p_h and p_d. The latter was, in addition, partitioned into the electrostatic, dispersion, and steric repulsion components, hereafter p_d^{elst}, p_d^{disp}, and p_d^{rep}, respectively. The simulation showed that none of the pressure components could be neglected at short separations. Among these non-negligible components was p_d^{elst} originating from direct electrostatic interaction between the lipid headgroups in the opposite membranes. Due to the high packing density of the membranes, the out-of-plane motion of the lipid molecules was restricted. In addition, the force field used involved a defect which forced the ethanolamine group to assume an unlikely cis conformation about the C –C bond. All this suppressed molecular and headgroup protrusions, so that no perceptible manifestation of the protrusion force could be detected. In the present study, we extend our GCMC simulations to a substantially larger area per molecule, A= 51.2 Å2 , as observed experimentally for the fluid-phase DLPE at 308 K.11,12 The simulations are carried out at three selected substrate-to-substrate separations corresponding to repulsive pressures on the order of 1 kbar 10^9 dyn cm⁻1. The distribution of the water-mediated pressure over its major components is analyzed against the water-induced changes in the membrane structure. A few comparative simulations are also made for the phosphatidylcholine PC analog of DLPE, DLPC.

METHOD

Although the ensemble averages $\langle f_{w \to m} \rangle$ and $\langle f_{m' \to m} \rangle$ needed to evaluate the intermembrane pressure can in principle be calculated by molecular dynamics MD—the predominant method used in studies of lipid membranes—we preferred to resort to the GCMC technique. This choice is motivated by, at least, two reasons. First, GCMC is best suited for simulating a SFA experiment, where confined water is allowed to exchange molecules and is in the chemical equilibrium with a bulk water reservoir Fig. 1. Unlike MD simulations, where the number of water molecules per lipid, n_w, is kept fixed at the respective experimental value, GCMC does not need the knowledge of n_w and treats it as a variable which adjusts itself so as to equilibrate confined water with bulk water. Second, with a judicious choice of the sampling procedure and parameters, GCMC is more efficient in exploring the configurational space of the hydrated membrane. Unlike MD, GCMC is not tied to the time evolution of the system, and so it can efficiently explore the membrane's con- figurational space using various "unphysical" moves. This is particularly important in simulations of hydrated lipid membranes where an adequate description of the permeation of FIG water in the membrane may involve difficulties. The difficulties can be appreciated from an estimate of the number of water molecules that cross the membrane in the accessible time scale of MD simulations.13,14 For a typical length of the MD trajectory 1 – 10 ns and a typical area of the membrane repeat unit $(\sim 1000 \text{ Å}^2)$ 2 , this number proves to be as small as 0.01–0.1. That is the probability of a water molecule finding its way to the membrane interior by normal diffusion is extremely low. By contrast, in a GCMC simulation, the water molecule need not diffuse through the whole monolayer thickness to reach the middle of the membrane. It can well do this by an unphysical particle insertion move. The stimulating role of unphysical moves in equilibrating the distribution of water through the membrane has been demonstrated by Jedlovszky and Mezei,14 who compared water density profiles calculated by GCMC and a standard NVT ensemble Monte Carlo. Note that the latter is similar to MD in the respect that in both cases water molecules may penetrate into the membrane only by diffusion. In similar conditions, GCMC showed a substantially deeper water penetration and an order of magnitude faster convergence to equilibrium. A. F

Force Field

Because of a high computational cost of GCMC simulations, where most of attempted configurations are spent for unsuccessful insertions, our choice of the inter- and intramolecular potential functions was restricted to computationally less demanding force fields of the united-atom type. As in our simulations at the crystal-state DLPE density,9 the conformational and intermolecular energies of DLPE molecules were calculated using an AMBER-based force field, whose torsional parameters were refined by Smondyrev and Berkowitz15 based on ab initio energy profiles and equilibrium geometries of eight compounds modeling individual fragments of dipalmitoylphosphatydylcholine DPPC. For the ethanolamine group, which is absent in DPPC, we initially used the relevant potentials from the GROMACS force field.16 Although the latter is a united-atom force field, it treats the hydrogen atoms of the amino group explicitly. Each hydrogen atom bears a partial charge but no LennardJones force center. The interactions of water molecules between themselves were described with the four-site TIP4P model.17 Trial simulations of hydrated DLPE membranes showed however that the earlier-described force field led to an unrealistic distribution of the O–C–C–N dihedral angles in the ethanolamine group, such that the most preferred conformer about the C –C bond was cis. To make sure that this result was not due to the interaction of the ethanolamine group with water, we carried out simulations of an anhydrous crystal-like DLPE bilayer. The most preferred conformer proved again to be cis, which was at variance with the available x-ray diffraction data for crystalline phospholipids,18 which favored a gauche conformer. An analysis of the simulated DLPE conformations showed that the cis conformer was stabilized by a strong N –H⁻O hydrogen bond, which formed between the ethanolamine oxygen and nitrogen at oms and whose formation led to large deformations of the bond angles in this group. In an attempt to avoid this hydrogen bond, we unsuccessfully tried the torsion O–C–C–N potential suggested by Smondyrev and Berkowitz15 for choline and claimed to be suitable for ethanolamine as well. Also unsuccessful was our attempt to solve the problem by placing Lennard-Jones force sites on the amine hydrogen atoms, as in the all-atom version of the AMBER force field.19 In the end, the undesirable hydrogen bonding was avoided by a twofold attenuation of the 1–5 Coulombic interaction H^-O in the ethanolamine group.

In an attempt to improve the balance of the water-lipid, water-water, and lipid-lipid interactions, we also tried to damp the water-lipid interactions by multiplying the relevant contributions to the instantaneous potential energy and force by a factor 1. The incentive to this part of our study will be clear from the discussion of the simulation results in the next section. In calculations of the potential energy of the DLPE-water system, the long-range contribution to the electrostatic energy was treated in the group-based dipole-dipole approximation. The DLPE molecule was represented as a set of electrically neutral groups as described by Smondyrev and Berkowitz.15 The reliability of the dipole-dipole approximation for long-range electrostatic interactions in DLPE membranes has been demonstrated elsewhere.9 Quite an acceptable accuracy of 0.3% for the electrostatic energy was achieved when the truncation radii for charge-charge and dipole-dipole interactions were taken to be 20 and 100 Å, respectively. Such an accuracy is noticeably worse than that reported for the particle mesh Ewald method usually used in MD simulations.20 It hardly makes sense, however, to strive for a better accuracy because the very representation of the molecular continuous charge distribution in terms of partial atomic charges introduces a much larger error in the electrostatic interaction energy.

As in our previous work,9 only the two inner DLPE monolayers were treated explicitly, whereas the outer monolayers were considered as generalized substrates for the inner ones Fig. 1. The associated substrate potential described, in a mean-field manner, the interaction between the adjacent inner and outer DLPE monolayers. It was parametrized so as to fit atomistic force field results for the interaction energy of two DLPE monolayers facing each other with their hydrophobic sides. The parametrization was made in the following way. We first calculated the non-bonded energy, Uz, of a perfect crystal-state DLPE bilayer as a function of the separation z between its constituent monolayers. The separation was defined as the spacing between the planes passing through the γ-chain methyl carbons in each monolayer. To bring the area per molecule to the fluid-phase value of 51.2 Å2 , the bilayer lattice periods were stretched by a factor of 51.2/ 38.6 1/2= 1.152. The monolayer-monolayer interaction energy was calculated using the above-mentioned AMBER-based force field.15 The position of the minimum of $U(z)$, z_m, proved to be close to zero separation $z_m = 0.15$ Å; i.e., in the minimum energy configuration the γ-chain methyl carbons of the two monolayers lay nearly in 42 Perts

Table I: Parameters of the atom-substrate potentials in Eq. 1

Atom	n	m	$C_n \times 10^{-3}$ (kcal mole^{-1} Å^n)	C_m (kcal mole^{-1} Å^m)	z_0 (Å)
Methylene carbon	10	4	1211	740	−3.64
Methyl carbon	10	4	1476	900	−3.64

he same plane, in agreement with the experimental crystal structure ($z=0.08$ Å).[10] For the potential well depth, we found $\varepsilon = -U(z_m) = 4.8$ kcal mole^{-1}. . The calculated dependence Uz was then fitted by the sum of atom-substrate potential functions so as to exactly reproduce z_m and ε and also to get the best agreement for the curvature of Uz at the minimum. The monolayer-substrate potential was of the

$$\Phi = \sum_i [C_n(z_i - z_0)^{-n} - C_m(z_i - z_0)^{-m}](n > m),$$

(1)

where the summation is over the carbon atoms of the hydrocarbon tails of the inner monolayer; m, n, C_n, C_m, and z_0 are the potential parameters to be found; z_i is the separation of atom i from the generalized substrate placed at the plane of the γ-chain methyl carbons of the outer monolayer. The final parameter set is presented in Table I.

It should be noted that in our previous simulations9 the substrate-monolayer interactions were described by another model potential, with $z_m = 3.5$ Å. The advantage of the present choice, with the minimum of Φ close to the crystalstate bilayer center, is that the substrate-to-substrate separation h is now closer to the lamellar repeat period, D, as measured by diffraction methods in multilamellar lipid dispersions. For the perfect crystal-state bilayers used in parametrization of Φ, the bilayer center is located at a distance of $z_m / 2$ from the substrate, so that h−D= 2zm / 2= 0.15 Å. For the gel and fluid states, the difference between D and h does not seem to be greater than 1 − 1.5 Å. Thus, in a MD simulation of the gel-phase DPPC, Venable et al.21 found that the average position of the γ-chain methyl carbons was 0.6 Å from the bilayer center. Would the generalized substrate model with parameters from Table I reproduce exactly the same density distribution, the substrate position would be ~0.15 Å Å from the average position of the γ-chain methyl carbons, so that the difference between D and h would be 20.6− 0.15= 0.9 Å. In actual fact, the difference may be somewhat larger because of inaccuracies involved in the mean-field treatment of the outer monolayers. For fluid phase DPPC, the MD result for the average deviation of the methyl carbons from the bilayer center is 2.06± 2.4 Å with no distinction between the β and γ chains. Considering that the chain is, on the average, closer to the bilayer center by about one C –C bond length (~ 1.5 Å),[22], 22 the average position of the γ-chain methyl, carbons relative to the bilayer center, as well as the difference between D and h, should not differ significantly from those for the gel phase.

Starting Configurations

The starting configurations of the system were constructed on the basis of the structure of DLPE monolayers observed experimentally in the DLPE-acetic acid crystal. The simulation cell contained a total of 64 lipid molecules 32 molecules per monolayer or 44 monolayer unit cells. The monolayers were placed with their hydrophobic sides on two parallel generalized substrates spaced a distance h. The initial separation of the terminal carbon atoms of from the respective generalized substrates was set at 0.15 Å, i.e., at the equilibrium distance of the monolayer-substrate potential $\Phi(z)$. The gap between the monolayers was first filled with water molecules at random positions and orientations, with the only constraint that the distance between two water oxygens and between a water oxygen and a DLPE atom was not less that 2.9 Å. After that a low temperature T=5 K NVTensemble Monte Carlo run was carried out to allow the system to assume an energetically favorable configuration. It is this configuration which was used in GCMC simulations as the starting one. C. Simu

Simulation Protocols

In our GCMC simulations only the number of water molecules was allowed to fluctuate, while the number of lipid molecules was fixed, i.e., the system was actually treated in a semigrand canonical ensemble.[23] To improve the efficiency of water insertions, the excluded volume mapping,[24] closely related to the Mezei's cavity-bias technique,[25] was employed, with the shortest allowed water-water separation of 2.4 Å. Further improvement in sampling efficiency was achieved using a Swendsen-Wang filtering.[26] The filtering out of energetically unfavorable insertions and deletions was based on a computationally inexpensive energy predictor, in which the electrostatic contribution to the water-water and water-lipid interactions was omitted. The frequencies of attempting water insertions and deletions were taken to be in the ratio 10:1.

A displacement move of a water molecule was made by translating its center of mass by a random vector and then rotating the molecule by a random angle about one of the three space-fixed axes chosen at random. A move of a DLPE molecule was defined to include a conformational change and a positional displacement of the molecule as a whole. The changes in the molecular conformation of DLPE were made subject to bond-length constraints, based on a rotational displacement algorithm suggested in our early work.27 A single conformational change affected all torsion and bond angles in the lipid molecule. The moves of water molecules were attempted at a ten times higher rate than the moves of lipid molecules.

The chemical potential of confined water was specified by setting the "density-corrected" excess chemical potential $\mu'' = \mu' + kT \ln d,$[28] where μ' is the excess configurational chemical potential29 and d is the average water density in grams per cubic centimeter. For bulk liquid water, d is close to unity and μ'' is practically equal to μ'. The value of μ''

TABLE II: Hydration degree and water-mediated pressure kbar as a function of substrate-to-substrate separation h and damping factor α.

h, Å	α	n_w	p	p_h	p_d	p_d^{elst}	p_d^{disp}	p_d^{rep}
42.6	1.0	11.0	5.0	4.1	0.9	0	−2.1	3.0
44.1	1.0	11.7	4.4	2.0	2.4	3.0	−0.7	0.1
	0.9	9.8	1.6	2.1	−0.5	−1.5	−2.0	3.0
	0.8	8.5	0.5	1.4	−0.9	−1.5	−1.8	2.4
	0.7	6.3	−0.8	1.4	−2.2	−6.7	−3.6	8.1
45.6	1.0	13.1	3.7	1.4	2.3	2.0	−0.6	0.9

was set at −5.95 kcal mole^{-1}, as determined in separate simulations of bulk TIP4P water so as to reproduce best the experimental density of water at T= 308 K. The three selected separations used in our simulations were h= 42.6, 44.1, and 45.6 Å. For each particular h tried, a series of sequential GCMC runs was carried out until the difference in <N> between consecutive runs was within 3–5 particles. The length of an individual run was 1.510^6 GCMC passes, each comprising N_0 moves, where N_0 is the initial number of water molecules in a given pass. The total number of con- figurations attempted at a given h amounted to 10^{10}.

RESULTS AND DISCUSSION

Force-Distance Dependence

We began our simulations with h= 44.1 Å, well below the experimental value of D at zero atmospheric pressure, i.e., at full hydration D^0= 46.1± 0.3 Å. 11,30

The starting number of water molecules generated at random as described in the previous section was 180, which corresponded to n_w = 2.8. After a series of eight GCMC runs, n_w stabilized at 11.7 Table II, the results for = 1. From the atomic density pro- files along the normal to the substrate plane Fig. 2, one can appreciate a deep penetration of water in the membranes, such that a perceptible water density occurs even in the hydrocarbon region. Several water molecules attained the hydrocarbon tails can also be observed in Fig. 3, which presents a snapshot of a typical configuration of the equilibrated system. In the context of the discussion of molecular protrusions, the positional disorder of the DLPE molecules along the z axis should be noted. This kind of disorder is well seen from the density profile of the C_2 atoms Fig. 2, which are the topological centers of the DLPE molecules. The extent of positional disorder can be quantified in terms of the width of the C_2 density profile, which is as large as 8.6 Å. That is, the protrusions of the individual lipid molecules from the average level into the aqueous phase may roughly be up to 4.3 Å, i.e., more than three CH_2 units.

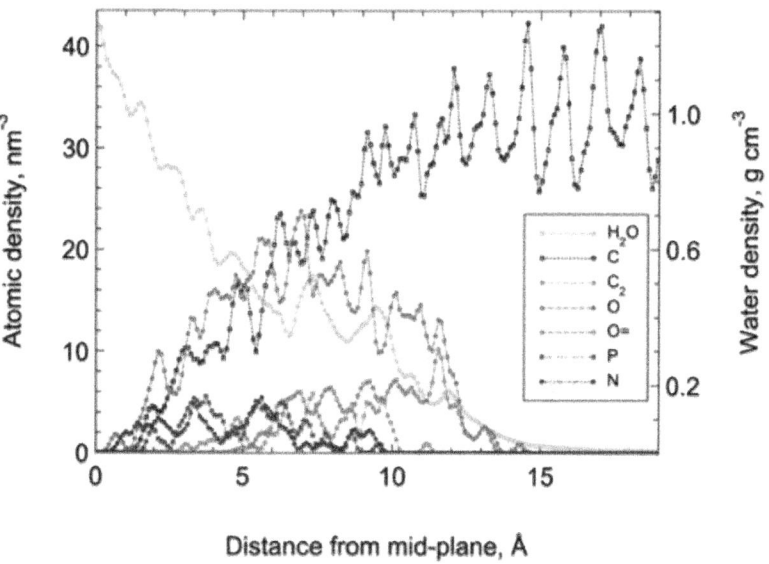

Figure 2: Atomic density profiles in hydrated DLPE membranes at h= 44.1 Å. To improve statistics, the profiles are symmetrized e.g., averaged over the two symmetrically equivalent parts of the system.

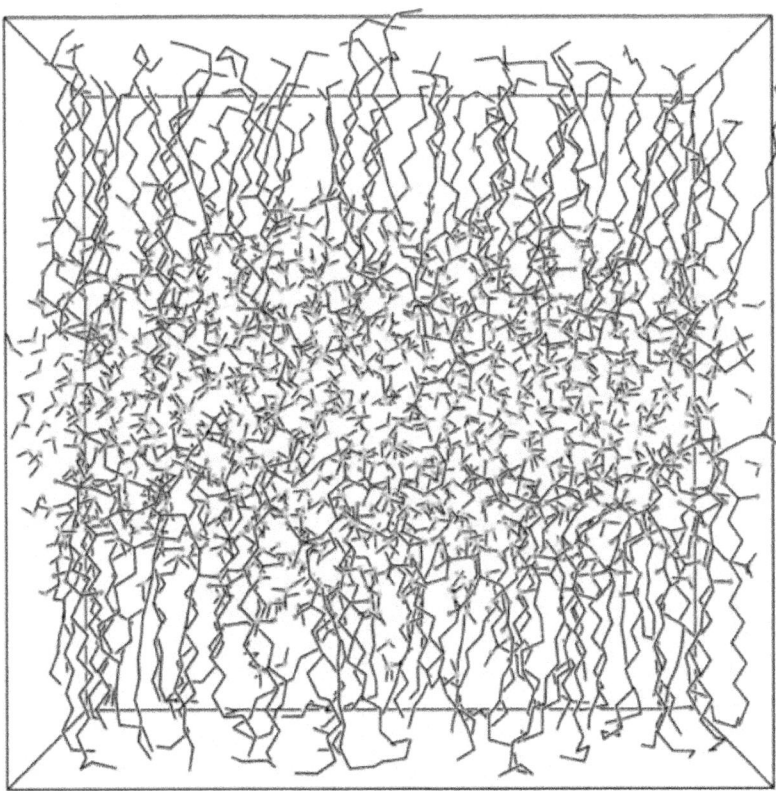

Figure 3: Snapshot of a typical configuration of the DLPE-water system at h= 44.1 Å. The DLPEs O, N, and P atoms are shown as red, blue, and purple spheres, respectively, whereas the water O atoms are represented by turquoise spheres. The carbon and hydrogen atoms are not shown.

drocarbon tails can also be observed in Fig. 3, which presents a snapshot of a typical configuration of the equilibrated system. In the context of the discussion of molecular protrusions, the positional disorder of the DLPE molecules along the z axis should be noted. This kind of disorder is well seen from the density profile of the C_2 atoms Fig. 2, which are the topological centers of the DLPE molecules. The extent of positional disorder can be quantified in terms of the width of the C_2 density profile, which is as large as 8.6 Å. That is, the protrusions of the individual lipid molecules from the average level into the aqueous phase may roughly be up to 4.3 Å, i.e., more than three CH_2 units.

As the system approached equilibrium and the number of water molecules increased, the phosphatidylethanolamine PE headgroup backbone, C_2 –C_1 –O_{11} –P–O_{12} –C_{11} –C_{12} –N,[31] experienced substantial conformational changes. In the initial crystal state, the conformations about the bonds O_{11} –P, P–O_{12}, and

C_{11} –C_{12} are fairly close to gauche to within $10°$, whereas the torsion angle describing rotation about the O_{12} –C_{11} bond is $106°$. As a consequence, the headgroup has a convoluted conformation, such that the $N^- C_2$ distance is 6.5 Å. With increasing N, the conformational distribution of the headgroups exhibited more extended conformations due to the appearance of additional, trans conformers about the bonds of the headgroup backbone. The extension of the headgroups is illustrated in Fig. 4 data for $= 1$, which depicts the ensemble averaged distribution of the $N^- C_2$ distances in the equilibrated system. The obvious driving force behind the unfolding of the strongly hydrophilic headgroup is the tendency to make its potential hydrogen bonding sites more accessible to water TABLE II

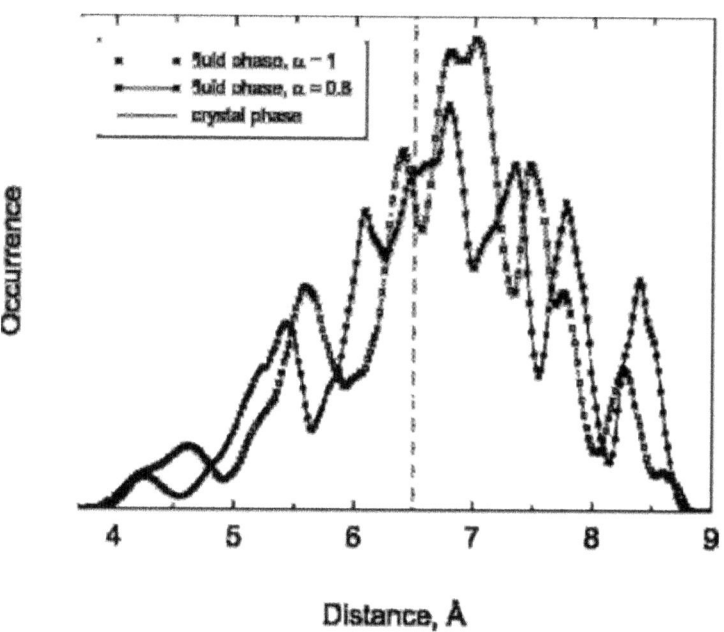

Figure 4: Distribution of the $N^- C2$ distance in hydrated DLPE membranes and DLPE-acetic acid crystal see Ref. 10.

molecules. Aside from becoming longer, some headgroups assumed orientations close to the normal to the membrane plane, n, directed towards the apposing membrane. This was well seen from an increase in the orientation parameter $S_{0.8}^{PN}$ which characterized the proportion of such orientations in terms of the cosine distribution of angles i formed by PN vectors with the respective normal n. More precisely, $S_{0.8}^{PN}$ was defined as the average probability density of $\cos(\theta_i)$ in the range $\cos(\theta_i) > 0.8$ $(\theta_i < 37°)$. The net result of the unfolding and reorientation of headgroups was the appearance of headgroups protruded into the aqueous phase. These headgroups can well be seen in Fig. 5, which shows the same configuration as in Fig. 3, except that the water molecules are not shown for clarity.

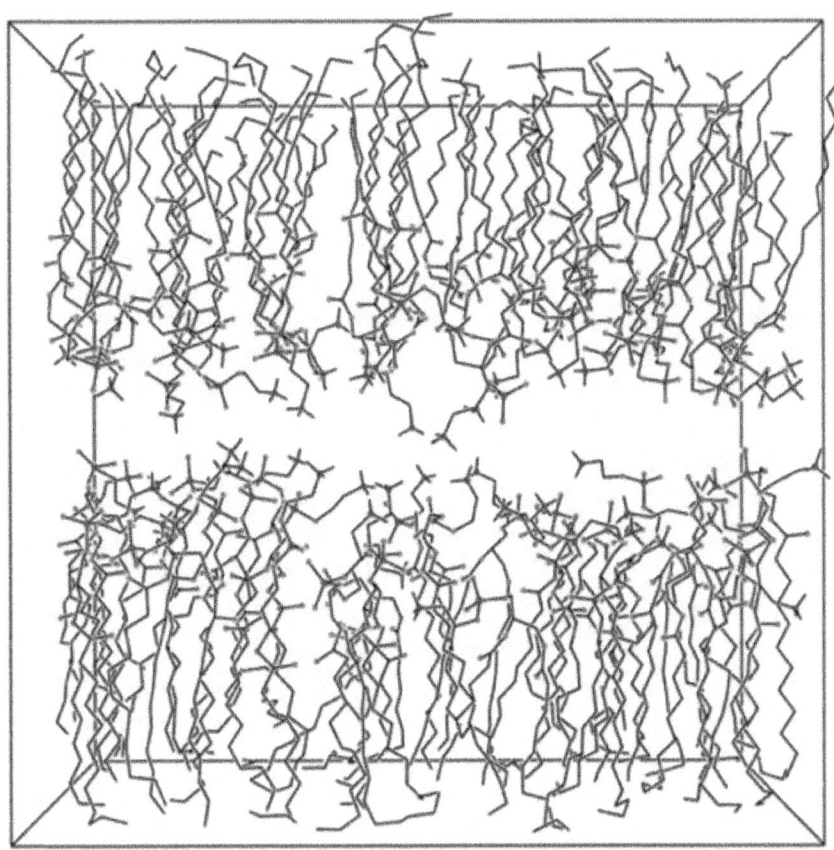

Figure 5: Same as in Fig. 3 but without water molecules.

As seen from Table II, at h= 44.1 Å the membranes experience a strong repulsion. A large part of this repulsion originates from the hydration pressure ph. However, the dominant contribution is due to the direct electrostatic interaction between the membranes, as described by p_d^{elst}. Moreover, nearly all pd elst comes from its static part, $p_d^{elst,0} = 2.9$ kbar. An analysis of the contribution of individual lipid pairs to $p_d^{elst,0}$ in the final configuration of the system shows that the repulsion arises mainly from the molecules whose headgroups are protruded towards each other from the opposite membranes. A pair of such molecules, which affords about 6% of $p_d^{elst,0}$, is shown in Fig. 6a. Although the interaction energy of these molecules is positive $(\sim 14 \text{ kcal mole}^{-1})$, it is negligibly small compared to the variable part of the total potential energy of the hydrated membranes $(-2.6 \times 10^5 \text{ kcal mole}^{-1})$. That is why the occurrence of such molecular configurations is not improbable. It is important to note that the upper of the molecules depicted in Fig. 6a showed the largest out-of plane displacement in the final configuration of the system $(\sim 3.8 \text{ Å})$. At the same time, a significant repulsive contribution to $p_d^{\text{elst},0}$ $(5\% - 6\%)$ was also observed for molecular pairs whose constituent molecules showed only slight out-of plane displacements. This suggests that the protrusions of lipid molecules as a whole do not play a decisive role in the short-range repulsion. The repulsion originates from the water-induced unfolding and reorientation of headgroups, leading to the appearance of repulsive configurations such as the one depicted in Fig. 6a.

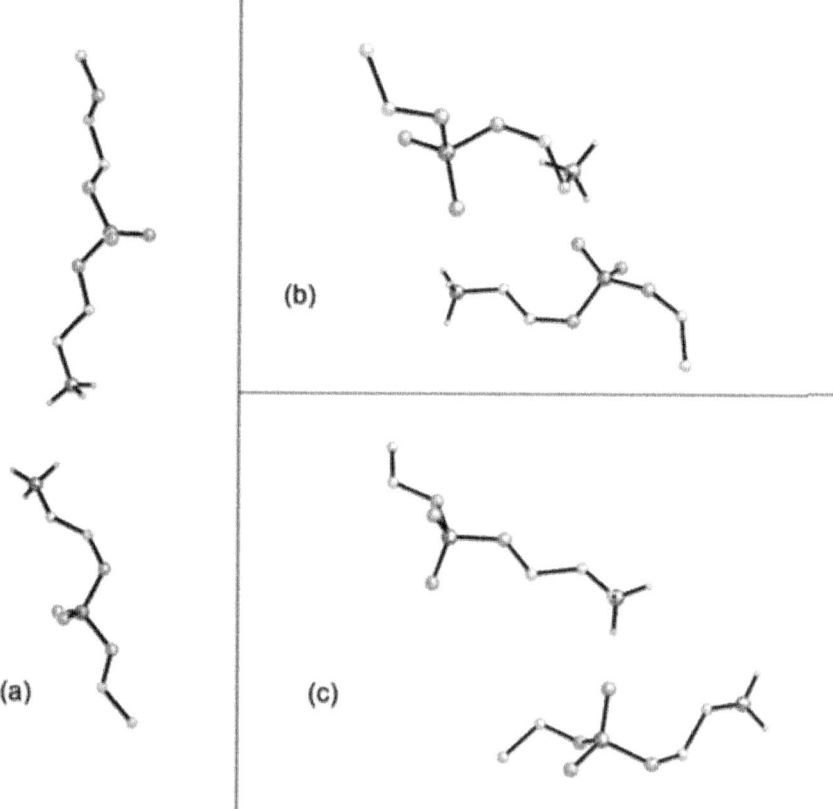

Figure 6: Typical pairs of opposing PE headgroups characterized by strong electrostatic repulsion a and attraction b, c. 45 P

The increase of the substrate-to-substrate separation to 45.6 Å led, as expected, to a drop in the water-mediated pressure, mainly at the expense of a decrease in its hydration component Table II. The electrostatic repulsion between protruded head groups remained the dominant contribution to the total pressure. Quite a different situation was observed at the shortest separation tried, h= 42.6 Å. Because of the closer approach of the membranes to each other, it became possible for the opposing head groups to assume energetically favorable configurations like the ones shown in Figs. 6b and 6c. In configuration b, the positively charged ammonium groups are located opposite the negatively charged phosphate groups, which results in a strong attractive contribution to p_d^{elst} . The characteristic feature of configuration c is the formation of a hydrogen bond N –H⁻O between the ammonium and phosphate group. The attractive configurations coexisted with repulsive ones such as in Fig. 6a, with the net result that p_d^{elst} proved to vanish. At this separation, the dominant pressure component was ph, although p_d^{disp} and p_d^{rep} were almost equally significant. Note that the occurrence of an attractive interaction between the phosphate group in one membrane and the ammonium group in the apposing membrane has been assumed by McIntosh and Simon32 in an attempt to explain the much smaller equilibrium intermembrane spacing and maximum hydration of PEs as compared to the respective PCs. The simulations show, however, that such an attraction may indeed exist only at very short separations and it vanishes as the separation is increased. To gain more confidence in this conclusion, we undertook two short GCMC runs at h= 44.1 and 45.6 Å starting from the final configuration of the system at h= 42.6 Å. The difference in thickness was initially compensated by inserting an empty space in the center of the water layer. After 10^6 GCMC passes, the electrostatic attraction fell off and p_d^{elst} rose to 2 – 3 kbar.

Hydration Degree

A disappointing result of the simulations is a too high hydration degree of the membranes. At the largest separation tried, h= 45.6 Å, the hydration degree $n_w = \langle N \rangle / 64 = 13.1$, which even exceeds the experiment-based estimate of maximum hydration, $n_w^0 = 8.8 - 10.2$.11,30 A consequence of the overestimated hydration is a shift of the region of strong repulsion to larger h. At h= 45.6 Å, the water-mediated pressure is still of the order of several kilobars Table II, whereas the experimental pressure at the respective D $(\sim 47$ Å$)$ should be close to zero.11,3

The too high hydration predicted by the simulations clearly points to a deficiency of the force field used, namely, to exaggerated hydrophilicity of the membranes due to an overestimated strength of the water-lipid interactions relative to the water-water and lipid-lipid ones. The water affinity of the DLPE membranes can be appreciated from Fig. 7, which shows the profile of the average interaction energy of a water molecule, u(z), with all its surroundings as a function of the separation of the molecule from the midplane of the system. Also shown are the contributions to u(z) from the waterwater and water-lipid interactions. The horizontal dot-and dash line at -19.8 kcal mole^{-1} indicates, as a reference, the average interaction energy of a water molecule with its surrounding water molecules in bulk water at 308 K. It can be seen that the residence of a water molecule in the hydrophilic region of the membranes is up to 7 kcal mole-1 more favorable in energy compared to the water bulk. Even in the vicinity of the mid-plane, the water-lipid contribution to u(z), which here comes mainly from protruded head groups, is noticeably greater in magnitude than the water-water contribution.

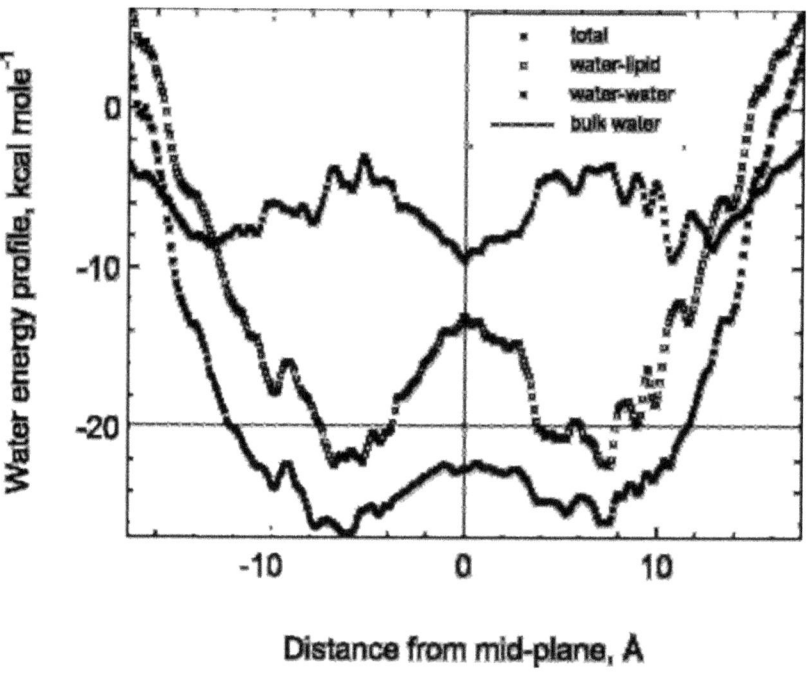

Figure 7: Profiles of the average interaction energy of a water molecule with its surroundings.

Effect of Methylation

In view of the property of the force field to exaggerate the hydrophilicity of DLPE, it was of interest to see whether the force field is capable of describing, at least qualitatively, the effect of full methylation of the terminal $-NH_3^+$ groups on hydration i.e., the effect of going from DLPE to DLPC. In the AMBER-based force field,[15] the methyls of the $-N(CH_3)_3^+$ group are treated in the united-atom approximation, so that the whole group is represented by a total of four force sites placed at the four nonhydrogen atoms. In this approximation, the $-N(CH_3)_3^+$ group differs from the $-N(CH_3)_3^+$ one only in two respects: 1 the N –C bond is by half longer than the N –H one and 2 each united $CH3$ atom of the former group bears both an electric charge 0.4 e and a van der Waals force site, whereas each H atom of the latter group bears the same charge but no van der Waals force site. As a consequence, the $-NH_3^+$ group is capable of forming hydrogen bonds with water molecules, while $-N(CH_3)_3^+$ group is not. This difference can be appreciated from Fig. 8, which compares the distribution of water density around N atoms for both groups. For the $-NH_3^+$ group, the main maximum of the water density occurs at a O⁻N distance of 2.8 Å, which is typical of the N –H⁻O hydrogen bond.

Considering the lower hydrophilicity of the $-N(CH_3)_3^+$ group, it may appear strange that DLPC shows a much higher maximum hydration, x $n_w^0 \approx 35$,[33] compared to DLPE. Indeed, the common experience in studies of water confined between solid surfaces shows that the stronger the surface liquid binding energy, the higher the average density of confined liquid and the stronger the solvation hydration repulsion.[34] To clear up this point, we first studied the methylation effect in the "pure" form, namely, we just replaced the –NH3 + groups by the $-N(CH_3)_3^+$ ones while leaving the geometrical parameters of the system A and h unchanged. At h= 44.1 Å, the result was a decrease in nw from 11.7 to 9.4 Å, i.e., the methylated membranes imbibed substantially less water. In the next step, we increased A to its experimental value for DLPC A= 68.7 Å2 , i.e., noticeably higher than that of DLPE and repeated the simulation. The final n_w was 19, i.e., much larger than the respective simulation result for DLPE 11.7. This provides a reasonable explanation for the much higher hydration degree of DLPC: Despite the lower hydrophilicity, DLPC membranes imbibe more water because of their lower areal density and, as a consequence, higher permeability. The latter fact can be appreciated from Fig. 9, which compares the water density profiles in DLPE and DLPC membranes. In the central part of the density

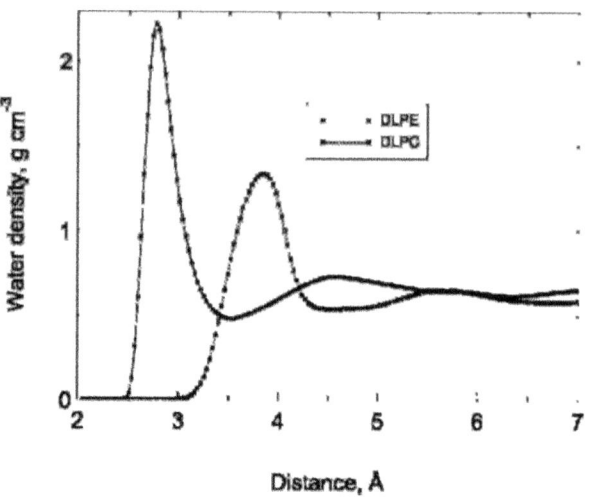

Figure 8: Water density distribution around N atoms in hydrated DLPE and DLPC membranes.

distributions $0 - 5$ Å from the midplane, the DLPC membranes show a lower water density because of the lower water affinity of the terminal $-N(CH_3)_3^+$ groups. However, on going toward the substrate, the water uptake by DLPC becomes higher than that of DLPE because DLPC provides more free space to accommodate water molecules. Unlike the case of DLPE membranes, where the available literature data on A, D, and n_w are restricted to their values at full hydration,[11,30] the structural parameters of DLPC membranes are known in a wide range of hydrations due to the x-ray diffraction study by Lis et al.[33] In comparing the experimental and simulation results, it should, however, be taken into account that our simulation and the measurements by Lis et al.[33] refer to different experimental conditions. The simulation mimics a SFA experiment, where the hydrated bilayers are compressed normally and the area per lipid A remains practically the same as that of the initially deposited bilayers. By contrast, the conditions of the experiments by Lis et al.[33] were equivalent to conditions of uniform hydrostatic compression, so that A noticeably decreased with decreasing n_w and D. The only reasonable comparison of the simulated and experimental systems is thus at similar values of A and D. Of the two values of A tried in our simulations A= 52.1 and 68.7 Å^2 , the former corresponds to a laterally compressed membrane and is more suitable for comparison. At h= 44.1 Å, which roughly corresponds to $D \approx 45.5$ Å,, the simulation resulted in nw= 9.4. The closest experimental values for A and D are 33 55 Å and 45 Å^2 , respectively, and the corresponding n_w= 6.2. That is the simulated hydration is again too high compared to the experimental value. D. S

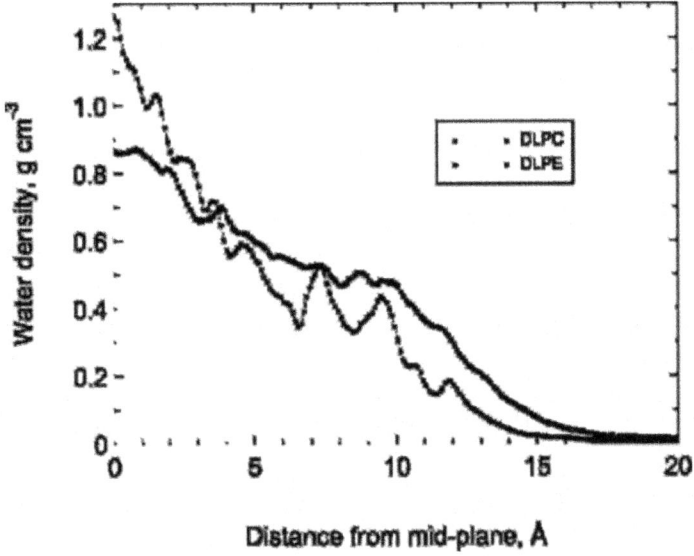

Figure 9: Symmetrized water density profile for hydrated DLPE and DLPC membranes at h= 44.1 Å

Simulations with Damped Water-Lipid Interactions

The discrepancy between the experimental and simulated hydrations is hardly surprising in view of the fact that the force field used, as well as other force fields accepted in simulations of hydrated lipid membranes,35 have never been tested for their ability to reproduce n_w in simulations of open systems, wherein water imbibed between the membranes is allowed to exchange particles with a bulk water reservoir.36 In most cases, the simulations were carried out using the NVT ensemble MD, with n_w, D, and A fixed at the respective experimental values. The experimental hydration level was also adopted in NPT ensemble simulations, where D and A were allowed to fluctuate so as to keep P close to atmospheric pressure. The use of the NPT ensemble did ensure the mechanical equilibrium for hydrated membranes but did not ensure the chemical equilibrium between the confined and bulk water. As far as the interfacial properties of water are concerned, the weakest point of the available united-atom force fields is, in our view, the parametrization of mixed water-solute interactions. In most cases, use is made of simple geometricand arithmetic-mean combining rules for parameters describing the interaction of force sites in dissimilar molecules. These rules are applied to both Coulombic and LennardJones potentials. In the former case, the geometric-mean combining rule means that the interaction between sites i and j in dissimilar molecules I and J is calculated with the same partial charges qi and q_j as used in describing the interactions of homomolecular pairs I   I and J   J. This rule would be strictly correct if q_i and q_j would be the actual physical charges on sites i and j. In fact, qi and qj are certain effective charges, which do not have a definite physical meaning and whose magnitudes strongly depend on the derivation method used.[37] In the particular force field[15] used in our simulations, the charges on the lipid molecules were derived from ab initio SCF 6-31G* level electron density by Mulliken population analysis.[38] By contrast, the magnitudes and positions of the charges on the water molecule were derived, together with the Lennard-Jones parameters of the water oxygen, so as to reproduce the experimental energy and density of water at ambient conditions. It would be surprising if the effective charges derived in so different ways could well describe the electrostatic water-lipid interactions. The situation with Lennard-Jones potential parameters is similar.

To appreciate the effect of the water-lipid interaction strength on the hydration level and water-mediated pressure, we undertook a series of GCMC simulations with damped water-lipid interactions. The basic results for = 0.9, 0.8, and 0.7 are presented in Table II. It can be seen that the damping of water-lipid interactions has a profound effect on the hydration level and pressure. Even a fairly small 10% damping = 0.9 reduces n_w by two water molecules per lipid and weakens the intermembrane repulsion by a factor of 2.6. In addition, the local water density at the midplane, which was as high as 1.25 g cm^{-3} at = 1 Fig. 2, reduces, at = 0.9, to about 1 g cm^{-3}. At a 20% damping = 0.8, one more molecule per lipid is lost and the repulsion drops by an order of magnitude, as compared to the initial force field. What is particularly interesting is the observed change in the mechanism of the water-mediated repulsion. Now the repulsion arises from the hydration component, whereas the net direct interaction between the membranes is attractive. The electrostatic component P_d^{elst} t also changes its sign and becomes attractive, as is typical of electrically neutral surfaces bearing laterally and orientationally mobile polar groups. The reason is the partial dehydration of the opposing headgroups, which increases their mobility and allows them to assume energetically more favorable configurations. An analysis of the structural changes occurring in the system with decreasing showed a reduction in the proportion of extended unfolded conformations of the headgroups. This change manifests itself in Fig. 4 as a noticeable drop in the occurrence of large N^-C_2 distances. In addition, the proportion of headgroups directed preferentially towards the opposing membrane decreased: The respective orientation parameter S0.8 PN fell from 0.8 to 0.6.

The change in the direct electrostatic intermembrane interaction from repulsion to attraction led to a substantial increase in the magnitude of the dispersion attraction P_d^{disp} and exchange repulsion P_d^{rep}. The reason is clear: With the initial force field = 1, the direct electrostatic repulsion of the membranes was longer ranged and it did not allow the opposing headgroups to approach close together. The change of the electrostatic repulsion to attraction allowed closer contacts of the headgroups, thus leading to an increase in the magnitude of P_d^{disp} and P_d^{rep}.

Further damping of the water-lipid interactions = 0.7 made the DLPE membranes hydrophobic, in the sense that the total water-mediated pressure became attractive Table II. The major factor responsible for the observed attraction was a strong attractive electrostatic component P_d^{elst} associated with lipid configurations similar to those in Figs. 6b and 6c. IV. CO

CONCLUSIONS

Our GCMC simulations of hydrated DLPE membranes using a united-atom AMBER-based force field have revealed a mechanism that may in principle be responsible for the short-range intermembrane repulsion. In this mechanism, the water-mediated repulsion originates from the direct electrostatic interaction of PE headgroups protruded towards each other from the opposing membranes. The driving force behind the headgroup protrusions is a strong water affinity of the headgroups, which favors headgroup unfolding and reorientation. That is the origin of the protrusions has to do rather with the energetic than entropic factor, as in the model suggested by Israelachvili and Wennerström.2,7 The relevance of this mechanism to real intermembrane repulsion is, however, unclear because the force field used fails to reproduce the hydration degree of the membranes. The simulations of the hydrated membranes with damped water-lipid interaction potentials have shown that the hydration level and watermediated pressure are very sensitive to the strength of the water-lipid interactions. Moreover, the damping of these interactions by only 10% – 20% changes qualitatively the role of the individual pressure components in the intermembrane repulsion. It is also found that the water-lipid interactions and the direct intermembrane interactions are involved in a complicated interplay through the effect of the former on the conformational distribution of lipid headgroups. All these findings calls for the development of a force field designed specially for simulation of biointerphases, with emphasis placed on an adequate description of the balance between the homo- and heteromolecular pair interactions and between the intra- and intermolecular contributions to the potential energy of the system. It is attractive, in particular, to try polarizable force fields for water and biomolecular systems,39 although the increased computational costs of such force fields over the fixed-charge models will hardly allow routine GCMC simulations of hydrated lipid membranes. Anyway, GCMC simulations can provide a stringent test for such force fields through a comparison of the experimental and simulated hydration levels.

Unlike theoretical treatments of the operative forces between neutral phospholipid membranes, where the direct electrostatic membrane-membrane interaction is usually ignored, the GCMC simulations reported in this article reveal a great importance of this kind of interaction, as represented by P_d^{elst} . Aside from a large relative magnitude of P_d^{elst} see Table II, it depends in a complicated way on hydration, membrane density, and intermembrane separation, thus being a controlling factor in the water-mediated interaction between the membranes.

REFERENCES

1.   R. Lipowsky and E. Sackmann, Structure and Dynamics of Membranes Elsevier, Amsterdam, 1995, Vol. 1.

2.   J. N. Israelachvili and H. Wennerström, J. Phys. Chem. 96, 520 1992.

3.   R. Evans and U. M. B. Marconi, J. Chem. Phys. 86, 7138 1987.

4.   I. Langmuir, J. Chem. Phys. 6, 873 1938.

5.   B. V. Derjaguin and N. V. Churaev, in Fluid Interfacial Phenomena,edited by C. A. Croxton Wiley, Chichester, 1986, Chap. 15.

6.   E. S. A. Jordine, J. Colloid Interface Sci. 45, 435 1973.

7.   J. Israelachvili and H. Wennerström, Nature London 379, 219 1996.

8.   V. A. Parsegian and R. P. Rand, Langmuir 7, 1299 1991.

9.   Pertsin, D. Platonov, and M. Grunze, J. Chem. Phys. 122, 244708 2005.

10. M. Elder, P. Hitchcock, and R. Mason, Proc. R. Soc. London, Ser. A 354,157 1977.

11. J. F. Nagle and M. C. Wiener, Biochim. Biophys. Acta 942, 1 1988.

12. Initially, we tried to simulate the interaction of gel-phase DLPE membranes at room temperature and A= 42 Å. It turned out, however, that the force field used failed to reproduce the gel state of the membranes, resulting in a substantially disordered structure corresponding rather to the liquid-crystalline state.

13. S.-J. Marrink and H. J. C. Berendsen, J. Phys. Chem. 98, 4155 1994.

14. P. Jedlovszky and M. Mezei, J. Chem. Phys. 111, 10770 1999.

15. M. Smondyrev and M. L. Berkowitz, J. Comput. Chem. 20, 531 1999.

16. D. P. Tieleman, URL http://moose.bio.ucalgary.ca/files/pope.itp.

17. W. L. Jorgensen and J. D. Madura, Mol. Phys. 56, 1381 1985.

18. H. Hauser, I. Pascher, R. H. Pearson, and S. Sundell, Biochim. Biophys. Acta 650, 21 1981.

19. W. D. Cornell, P. Cieplak, C. I. Bayly, I. R. Gould, K. M. Merz Jr., D. M.Ferguson, D. C. Spellmeyer, T. Fox, J. W. Caldwell, and P. A. Kollman, J.Am. Chem. Soc. 117, 5179 1995.

20. T. Darden, D. York, and L. Pedersen, J. Chem. Phys. 98, 10089 1993.

21. R. M. Venable, B. R. Brooks, and R. W. Pastor, J. Chem. Phys. 112, 4822 2000.

22. W. Shinoda, N. Namiki, and S. Okazaki, J. Chem. Phys. 106, 57311994.

23. The use of a statistical ensemble with a fixed lateral area of the simulation cell and a fixed number of lipid molecules could, in principle, partially suppress fluctuations in the local areal density of the lipid, which could affect the membrane permeability. For a simulation cell containing 32 symmetrically independent lipid molecules per monolayer, as in our case, the use of a fixed membrane area ensemble should however have little effect on the fluctuations in the local areal density. This follows from the MD results reported by S. E. Feller and R. W. Pastor, J. Chem. Phys. 111, 1281 2005 who compared the probability distributions of single molecule areas in fixed- and variable-area ensembles for a hydrated lipid membrane with 36 independent molecules per monolayer in the simulation cell.

24. M. R. Stapleton and A. Panagiotopoulos, J. Chem. Phys. 92, 1285 1990

25. M. Mezei, Mol. Phys. 40, 901 1980.

26. R. H. Swendsen and J.-S. Wang, Phys. Rev. Lett. 58, 86 1987.

27. A. J. Pertsin, J. Hahn, and H.-P. Grossmann, J. Comput. Chem. 15, 1121 1994

28. . A. Pertsin and M. Grunze, J. Phys. Chem. B 108, 16533 2004.

29. D. J. Adams, Mol. Phys. 28, 1241 1974.

30. T. J. McIntosh and S. A. Simon, Biochemistry 25, 4948 1986.

31. Throughout the article, the atom numbering is as suggested by Sundaralingam Ann. N.Y. Acad. Sci. 195, 324 1972. The two carbon atoms in parentheses belong to the glycerol residue.

32. T. J. McIntosh and S. A. Simon, Langmuir 12, 1622 1996.

33. L. J. Lis, M. McAlister, N. Fuller, R. P. Rand, and V. A. Parsegian, Biophys. J. 37, 657 1982.

34. We here imply a large separation range, where the oscillations of water density and hydration pressure can be neglected.

35. M. Schlenkrich, J. Brickmann, A. D. MacKerell, Jr., and M. Karplus, in Biological Membranes: A Molecular Perspective from Computation and Experiment, edited by K. M. Merz, Jr. and B. Roux Birkhäuser, Boston, 1996.

36. The GCMC simulations by Jedlovszky and Mezey see Ref. 14, cited in the beginning of the previous section, were carried out at a lamellar repeat period D much larger than D_0. No attempt to reproduce the experimental value of nw was undertaken.

37. J. Pertsin and A. I. Kitaigorodsky, The Atom-Atom Potential Method Springer, Berlin, 1987.

38. S.-W. Chiu, M. Clark, V. Balaji, S. Subramaniam, H. L. Scott, and E.Jakobsson, Biophys. J. 69, 1230 1995.

39. W. L. Jorgensen and J. Tirado-Rives, Proc. Natl. Acad. Sci. U.S.A. 102,6665 2005.

Chapter 2

COUPLING THE USE OF COMPUTER CHEMICAL SPECIATION MODELS AND CULTURE TECHNIQUES IN LABORATORY INVESTIGATIONS OF TRACE METAL TOXICITY

Michael R. Twiss[a], Olivier Errécalde[b], Claude Fortin[c], Peter G. C. Campbell[d*], Catherine Jumarie[e], Francine Denizeau[f], Edward Berkelaar[g], Beverley Hale[g] & Ken van Rees[h]

[a]Department of Chemistry, Biology and Chemical Engineering, Ryerson Polytechnic University, 350 Victoria Street, Toronto, Ontario M5B 2K3, Canada

[b]Les Laboratoires Aeterna Inc., 456 rue Marconi, Ste-Foy, Québec G1N 4A8, Canada

[c]Département de Protection de l'Environnement, Institut de Protection et de Sureté nucléaire, BP No 1, 13108 Saint-Paul-lez-Durance, Cedex, France

[d]INRS-Eau, Université du Québec, C.P. 7500, Ste-Foy, Québec G1V 4C7, Canada

[e]Département des Sciences biologiques, Université du Québec à Montréal, C.P. 8888, Succursale Centre-ville, Montréal, Québec H3C 3P8, Canada

[f]Département de chimie, Université du Québec à Montréal, C.P. 8888, Succursale Centre-ville, Montréal, Québec H3C 3P8, Canada

[g]Department of Land Resource Science, University of Guelph, Guelph, Ontario N1G 2W1, Canada

[h]Department of Soil Science, University of Saskatchewan, Saskatoon, Saskatchewan S7N 5B3, Canada

ABSTRACT

The bioavailability and toxicity of a dissolved metal are closely linked to the metal's chemical speciation in solution. A variety of inorganic and organic ligands are often used in laboratory toxicity tests to control the concentration of labile trace metal in solution. Computerised chemical speciation models based on thermodynamic principles can be used to estimate metal speciation under such experimental conditions. However, these models are sensitive to the quality of their thermodynamic databases. Detailed protocols for the incorporation of reliable equilibrium formation constants into widely available computer chemical speciation programs (e.g., MINEQL+ and MINTEQ) are provided. The examples demonstrate both the benefits and the potential pitfalls involved in the use of chemical speciation models. The application of chemical speciation modelling to metal toxicity studies is discussed and guidelines are proposed for its proper use. Both defined media and chemical speciation programs have co-existed for two decades but the combined use of these techniques has been reserved for those possessing in-depth knowledge of both chemistry and biology. The techniques presented should enable an investigator with basic biological, chemical and computing skills to design an aqueous medium and incorporate correct thermodynamic constants into a computer chemical speciation program, starting from a standardised database, thereby providing a sound framework for critically assessing the biological response of a particular test organism to a given metal.

INTRODUCTION

When environmental chemists use the term "speciation" with reference to metals, they generally mean the distribution of the metal in question among various possible forms or "species". It is often useful to distinguish between the physical speciation of a metal, i.e. its distribution among dissolved, colloidal or particulate forms, and the chemical speciation of the metal. This latter term is generally taken to mean the distribution of the metal among various distinct chemical species in solution, and it includes both the distinction between "complexed" and "uncomplexed" metals (see Figure 1), and the distinction between different oxidation states. Examples of different metal species or forms are given in Table 1; useful reviews (Honeyman and Santschi, 1988) and texts (Buffle, 1988, Morel and Hering, 1993, Stumm and Morgan, 1996) are available in the area of metal speciation.

Metal speciation is of importance in both the geochemical and toxicological areas. In the present context, however, we shall only be concerned with the influence of speciation on the bioavailability and toxicity of metals. Furthermore, in considering the interactions of trace metals with living organisms, one can identify three broad areas of concern: (1) metal speciation in the external environment; (2) metal interactions with the biological membrane separating the organism from its environment; (3) metal partitioning within the organism, and the attendant biological effects (Sunda, 1991). We have consciously limited the scope of this report to the first of these areas, and we have chosen to focus primarily on dissolved metals (rather than those associated with soil solids, aquatic sediments or ingested food items).

For 15+ years it has been widely recognised that the total aqueous concentration of a metal is not a good predictor of its "bioavailability", i.e. that the metal's speciation will greatly affect its availability to aquatic organisms (Campbell, 1995). It follows that metal speciation must be considered in the design and interpretation of experiments intended to evaluate metal bioavailability and toxicity. Though to some readers this statement may seem superfluous or unnecessary, nevertheless there are still abundant recent examples in the scientific literature of poorly designed experiments on metal-organism interactions and faulty interpretations, where changes in biological response are interpreted without due consideration of the role of metal speciation.

The purpose of this review is not to try to convince the reader of the importance of metal speciation (e.g. Tessier and Turner, 1995), but rather to identify some of the speciation-related pitfalls commonly encountered when designing experiments with metals and to provide practical guidelines on how to avoid these pitfalls.

STUDIES OF METAL-ORGANISM INTERACTIONS

Overview

In determining the toxicity and availability of metals to target organisms, it is critically important to consider metal speciation in the exposure media (Campbell, 1995). In order to calculate and thus control this speciation, the composition of the exposure media must be known as accurately as possible. This requirement is often met by using chemically defined culture media and computerised chemical equilibrium programs that enable rapid and precise estimates of chemical speciation in aqueous media based on thermodynamic principles. Much of the pioneering work on metal-organism interactions has been done with unicellular algae. Several culture media have been developed expressly for testing metal interactions with phytoplankton: e.g. the freshwater medium Fraquil (Morel et al., 1975) and the seawater medium Aquil (Morel et al., 1975, Morel et al., 1979, Price et al., 1991). In the design and use of these media, three main constraints were defined: (1) the need to limit inadvertent contamination of the nutrient salts and the water used to prepare the exposure media (i.e. proper choice of techniques, vessels, apparatus and reagents); (2) the need to avoid the formation of solid precipitates in the exposure media (e.g. by using low concentrations of major nutrients and trace metals, closer to natural conditions, thus ensuring that the solubility limits of the solid phases are not exceeded); and (3) the need to be able to solve the chemical speciation calculations for the exposure media. Although these constraints were specifically identified for studies with phytoplankton, they also apply to studies on other organisms subject to the aqueous exposure route (e.g. aquatic and terrestrial plants; aquatic animals). Knowledge of these constraints is imperative for researchers who wish to develop new bioassay media or to adapt classic culture media to their particular experimental conditions. As implied by constraints (1) and (2), nominal metal concentrations, i.e. those calculated on the basis of the dilution of a stock metal solution into the exposure medium, may differ considerably from the true values to which the test organism is exposed. To control for dilution errors, inadvertent contamination or metal losses from solution by sorption or precipitation reactions, metal levels should be measured in the experimental solutions, at the beginning and end of the exposure period.

With reference to constraint (3), the accuracy of results obtained from a chemical speciation calculation will depend on the validity of the equilibrium constants used. Ideally, the calculated concentrations of the various complexes in solution should represent what truly exists in the medium and be verified using

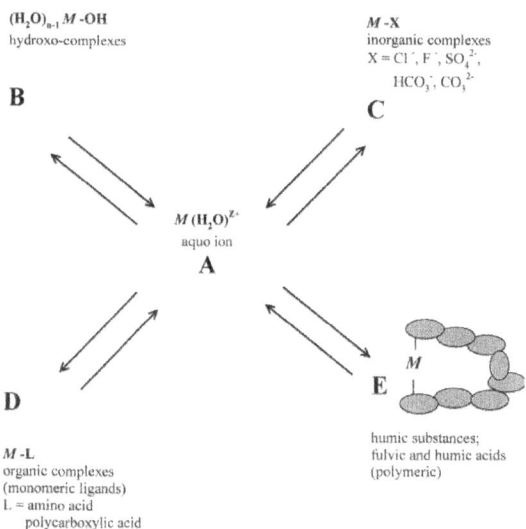

Figure 1: General scheme showing metal speciation in solution. Chemical equilibrium modelling can be used with confidence to calculate the distribution of metal, M , among forms A, B, C and D. Refer to Table 1 for environmentally significant examples of each of metal species.

analytical chemical techniques. However, it must be acknowledged that an extremely low level of a specific chemical form of a trace metal may be one factor that controls the biological response. For example, the free ferric aquo-ion, Fe^{3+} 6 H_2O, induces iron limitation in cyanobacteria at levels less than

10^{-19} mol L–1 (Wilhelm, 1995), a concentration that is currently beyond the ability of measurement using direct chemical analysis. We are therefore faced with the position of inferring chemical speciation based on thermodynamic principles, and hence, a heavy reliance on the integrity of the thermodynamic data used to calculate this speciation.

In the following sections, we describe a detailed approach to coupling the techniques of toxicity testing in defined chemical media with the accurate calculation of chemical speciation using a computerised chemical speciation program. The procedures described below should enable an investigator with basic biological, chemical and computing skills to design an aqueous medium and incorporate correct thermodynamic constants into a computer chemical speciation program starting from a standardised database. Such an approach provides a sound framework for critically assessing the biological response of a particular test organism to a given metal. In this discussion we emphasise the chemical speciation program MINEQL+, primarily because its user-friendly format allows for the simple modification of the thermodynamic database, and for easy manipulation of the output data from the modelling calculations.

Table : Examples of metal forms or species that are of potential environmental and toxicological importance

Metal form (A → E in Figure 1)	Examples
A. Free metal ion	$Al^{3+}(H_2O)_6$
	$Cu^{2+}(H_2O)_6$
B. Hydroxo-metal complexes	$AlOH^{2+}$, $Al(OH)_2^{1+}$, $Al(OH)_4^{1-}$
	$FeOH^{2+}$, $Fe(OH)_2^{1+}$, $Fe(OH)_4^{1-}$
	$Cu(OH)_2^0$
C. Simple inorganic complexes	AlF^{2+}, AlF_2^{1+}
	$CdCl^{1+}$, $CdCl_2^0$, $CdCl_3^{1-}$
	$HgCl_2$, $HgOHCl^0$
	$CuCO_3^0$
	$CdSO_4^0$
D. Simple organic complexes:	
i) synthetic	$Cu\text{-}EDTA^{2-}$
	$Cd\text{-}NTA^{1-}$
ii) natural	Cd-alanine
	Cd-citrate
	Fe-siderophore
E. Polymeric organic complexes	Al, Fe, Cu, Pb or Hg – fulvic
	or humic acid

TYPES OF EXPERIMENTS

The experiments of interest generally fall into two categories: (1) exposure experiments designed to determine metal uptake under a variety of conditions, and (2) biological tests in which a metal-induced response (either stimulatory or inhibitory) is followed as a function of metal concentration or some other experimental variable

Typically such experiments attempt to determine how the "biological response" (be it metal uptake per se or a metal-induced response) varies as a function of particular chemical or biological factors (Bryan, 1971; Langston and Spence, 1995). "Chemical" factors might include metal concentration (e.g. total, free-metal ion, some particular species) or other exposure conditions (e.g. pH, temperature, hardness, ionic strength, concentrations of dissolved organic matter or other potentially competing metals). "Biological" factors might include such variables as the choice of test species or strain, the biological life stage of the test species, or the antecedent conditions (culture conditions, prior exposure regime). In all cases, when it comes to interpreting the results of the experiment, the investigator must be able to distinguish between biological changes induced by the variable being manipulated and chemical speciation changes that may be indirectly affecting the biological response. To carry out metal-organism experiments of this type, i.e. in synthetic solutions, one must necessarily add the metal of interest to the exposure solutions. The fate of this added metal is of crucial importance. The investigator must ascertain if:

(i)   all of the added metal remains in solution (or is some metal lost from solution by precipitation of some insoluble solid phase, by adsorption onto the container walls or onto suspended solids, or by volatilisation?)

(ii)   the metal in solution remains as the free-metal (aquo) ion, $Mz+$? (or does the added metal undergo complexation reactions with the inorganic and organic ligands in solution?)

(iii)   the metal remains in the oxidation state (valence) in which it was added (or does the metal undergo oxidation-reduction reactions, notably under the influence of micro-organisms and light?)

(iv)   the metal is subject to alkylation/dealkylation reactions under the conditions of the bioassay.

Only if investigators can answer all these questions convincingly will they be able to interpret the results of the experiment unambiguously.

DESIGN OF THE ASSAY CONDITIONS

Composition of the culture medium: The composition of the basic exposure medium is of primary importance since it must be of known and definable chemical composition if the investigator is to be able to calculate the chemical speciation of the metal. Media containing only inorganic salts are close to ideal from the chemical point of view, but obviously not all test organisms will tolerate or survive in such systems. Photoautotrophs (e.g., phytoplankton, macroalgae, rooted aquatic and terrestrial plants) probably represent the least complicated test organisms from this point of view, whereas heterotrophic micro-organisms and animal cell cultures, with their requirements for complex and often exotic organic "supplements", pose serious problems for the design of controlled metal exposure conditions.

Even with photoautotrophs, however, the "standard" inorganic growth media may be unsuitable as exposure media, due to high pH, high chelator concentrations (added to keep essential metals in solution), or high concentrations of certain macronutrients (especially phosphorus), which may lead to precipitation of the added metal and loss from solution. Possible solutions in such cases include: (1) reducing the macronutrient levels in the exposure media; (2) removing the offending macronutrient from the exposure medium (and shortening the exposure period); and (3) pre-conditioning the test cells in the presence of high levels of the macronutrient (and then exposing these nutrientreplete cells in a nutrient-free exposure medium). As an example of this latter approach, pre-exposure of phytoplankton cells to high levels of phosphate allows the cells to accumulate intracellular polyphosphate that they can subsequently use as a source of phosphorus over several days growth in phosphate-free exposure medium (Kuwabara, 1985; Parent and Campbell, 1994). Similarly, higher plants may be maintained for several days in iron-free rooting solution, provided that they have accumulated iron in their biomass during a pre-exposure acclimation period. Note, however, that changing the nutrient status of phytoplankton cells may alter their sensitivity to metals; e.g., phosphorus-defi- cient Scenedesmus were more sensitive to copper than were more phosphorus-replete cells (Twiss and Nalewajko, 1992).

For test organisms and cells that normally require organic supplements (e.g., glucose, amino acids; peptone; serum; nutrient broth), it will usually be necessary to grow/maintain the cells in their normal medium, then harvest the cells (with a careful rinsing step to remove adhering medium) and re-suspend them in a minimal exposure medium that does not contain the organic supplement. However, this procedure cannot be applied in the case of adhesive, epithelial cells, since they lose their functional polarity if the confluent cells are resuspended; in such cases, the cells must be washed carefully while still on their culture support, and then covered with the supplement-free exposure medium. Exposure times will presumably have to be short (including verification of the viability of the cells and their energy status at the end of the control exposures in metal-free solutions).

Metal-complexing ligand: As described above, control of metal speciation in the exposure medium at the beginning of an experimental treatment is essential, but maintaining this chemical speciation during the metalexposure time is also required. The addition of metalcomplexing ligands that act as metal buffers is one way to ensure the constancy of the metal concentrations in solution (other complementary approaches include shortening the exposure times, and increasing the volume of the exposure solution relative to the biovolume). Ethylenediaminetetraacetic acid (EDTA) and nitrilotriacetic acid (NTA) are often used as metal buffers because they form hydrophilic complexes with metals that are virtually impermeable to biological membranes (Simkiss and Taylor, 1995). Ligands of choice must have well-established stability constants for their reactions with all cations present in the exposure medium, must be stable (i.e., not subject to chemical hydrolysis, redox reactions or to bio-degradation or photo-degradation) and must have a sufficient buffering capacity towards the metal under study

Among commonly used ligands, there are two kinds to avoid: (1) ligands forming lipophilic metal complexes (e.g., diethyldithiocarbamate, ethyl xanthate) and (2) low molecular weight organic metabolites (e.g., citrate, glycine). Ligand-metal complexes of these types are potentially able to cross cell membranes, leading to an increase in toxicity above that expected on the basis of the free-ion concentration (Poldoski, 1979; Block and Pärt, 1986; Guy and Kean, 1980; Pärt and Wikmark, 1984; Daly et al., 1990; Errécalde et al., 1998; Errécalde and Campbell, 2000).

Generally, biological responses to metals are tested along a range or gradient of free-metal ion concentrations, $[M^{z+}]$. This gradient may be achieved in a number of ways: (1) by varying the total concentration of the metal while keeping the metal buffer concentration constant; (2) by varying the total concentration of the metal buffer while keeping the total metal concentration constant; or (3) by varying the nature of the ligand.

In each case where the nature or the concentration of the metal buffer is manipulated, one can adjust the freemetal concentration over the desired range, but by doing so the free-ion concentrations of the other trace elements chelated by the metal buffer ligand are also changed. Among these elements, Ca^{2+}, Mg^{2+}, and other cationic trace metals can potentially compete with the metal of primary interest for uptake sites on the surface of the test organism (e.g. Cu^{2+}/Mn^{2+}: algal cell [Sunda and Huntsman, 1983]; Cu^{2+}/Ca^{2+}:: algal cell [Xue et al., 1988]; $Cd^{2+}/Ca^{2+} Cu^{2+}/Mn^{2+}$: fish gill [Pärt et al., 1985]). Given these interactive effects, it is essential to control not only the speciation of the metal of interest, but also the free-ion concentrations of other cationic species likely to compete with the metal of primary interest (e.g. for ligands in solutions, and for metal-binding sites at the biological surface); ideally the free-ion concentrations of these other cationic species should be held constant as $[M^{z+}]$ is varied.

The free ionic concentrations of other cationic species are less affected by varying the total concentration of the metal of interest than by manipulating the concentration or the type of the ligand. It is thus preferable to achieve a gradient of free-metal concentrations using a single ligand at constant concentration.

pH: pH is a key variable affecting chemical speciation (Stumm and Morgan, 1996). Changes in medium acidity, particularly in weakly-buffered freshwater media, may result from phytoplankton uptake of CO_2, NH_4^+ − (Nichols, 1973). Higher plants may also influence the pH of their rooting medium by excretion of low molecular weight organic acids and by differential uptake of NO_3^-/NH_4^+; this latter pattern has been linked to differential tolerance of wheat cultivars to Al (Taylor and Foy, 1985). Fish may affect the pH of their exposure medium by excretion of NH_4^+ or urea, or by respiratory release of CO_2 (Lin and Randall, 1990; Randall et al., 1991; Playle and Wood, 1989). Animal cells in culture may also affect the pH of their medium, through excretion of (acidic) metabolic products.

As was described for metal concentrations, using short incubation times and low biovolume:medium ratios may reduce variations in pH. If pH variations are >0.2 pH units during the incubation time then the use of a pH buffer is recommended. The choice of an appropriate pH buffer is not always straightforward, however.

The buffers N-2-hydroxyethyl-piperazine-N-2'- ethanesulfonic acid (HEPES: useful in the pH range 6.8–8.2) and piperazine-N,N'-bis(2-ethanesulfonic) acid (PIPES: useful in the pH range 6.1–7.5) are reported to be non metal-complexing (Good et al., 1966), but most of the supporting references seem to be qualitative (e.g., "undetectable complexation") and thus are inappropriate for introduction into computer models. Evidence of surfactant activity caused by HEPES and other zwitterionic buffers, which could enhance metal toxicity or interfere with analytical measurements, has been reported (Vasconcelos et al., 1996, Vasconcelos and Almeida, 2000). Soares et al. (1999) report that pH buffers of the n-substituted aminosulfonic acids with a morpholinic ring type (e.g. MES, 2-(N-morpholino)-ethanesulphonic acid; MOPS, MOPSO) do not interfere with the voltammetric determination of Cd and Pb: presumably other metals would also not be complexed by these buffers. The buffer tris(hydroxymethyl)aminomethane (TRIS) has in some cases been used to control both pH and free-metal ion concentrations, e.g. Cu^{2+} (see Sunda and Lewis, 1978), but some of the TRIS-M^{z+} formation constants are unavailable for low ionic strength (I = 0) and thus are difficult to apply under conditions that differ from those used for the original measurements. Also, some pH buffers and trace metal chelators may be toxic in their own right (Ferguson et al., 1980; Price et al., 1991; Lage et al., 1996).

Precipitation and Sorptive loss of Metal from Solution

Minimisation of precipitation and adsorption is an important part of the preparation of chemically defined media. These two phenomena are linked because ions in solution may be strongly adsorbed onto precipitates. Adsorption onto the wall of the medium container can be avoided by the choice of appropriate material such as polycarbonate, polytetra-fluoroethylene (Teflon, PTFE; the less costly high density polyethylene (HDPE) polymer can be an alternative to Teflon), or silanised glass (SurCote, Fisher Scientific).

Adjusting the ionic composition of the medium usually controls the formation of solids. Indeed, many media recommended for use in the study of trace metal interactions with phytoplankton are designed to avoid precipitation (see Price et al. (1991) for a detailed description of media preparation). Nevertheless, care must be taken to avoid the precipitation of solids during the sterilisation process. Sterilisation by filtration (0.2 m pore size polycarbonate filter) or microwave sterilisation (Keller et al., 1988; Price et al., 1991) is recommended.

Metallic Contamination

All the equipment intended to come into contact with solutions should be previously acid-washed using established protocols (cf. Nriagu et al., 1993). The washed equipment must be dried in a clean room or a laminar flow hood to avoid contamination resulting from the deposition of ambient dust, which is known to contain trace metals such as Cd and Pb (e.g., Feng and Barratt, 1994). For a compilation of the trace metal concentrations present in common laboratory materials (e.g., filters, paper tissues, powdered gloves), see Hume (1973) and Nriagu et al. (1993). A major source of trace metal contamination can often be the salts used to prepare the exposure medium. This is most relevant for the preparation of saltwater media that require upwards of 30 g of salt per litre, as opposed to freshwater media requiring significantly less (<0.3 g L–1). The highest quality reagent salts should be used wherever possible. The use of cation exchange resins to remove trace metal impurities from culture media is a very effective technique; Price et al. (1991) provide a detailed description of this approach. Alternatively, electroactive metals such as Cd, Cu, Pb and Zn can be removed from concentrated stock solutions by prolonged electrolytic reduction in systems containing a pool of mercury as the reducing electrode (the cations are reduced to their parent metal, which then forms an amalgam within the mercury pool).

CALCULATING CHEMICAL SPECIATION USING A COMPUTERISED MODEL

There exist several computer programs that perform chemical speciation calculations rapidly (e.g., MINEQL, MINTEQA2, PHREEQ, GEOCHEM). However, it is essential that a critically reviewed database of thermodynamic values be used for these calculations, rather than the default values supplied with the chemical speciation software; errors have been reported in the default databases for the MINTEQA2 and MINEQL programs (EPA, 1995; Serkiz et al., 1996). The main sources of error in the equilibrium constants are: (1) the use of equilibrium constants instead of formation constants, (2) formation constants not expressed at infinite dilution, and (3) poorly chosen constants. These sources of error are discussed in detail below.

The chemical speciation model used here is MINEQL+ (Version 3.01), available from Environmental Research Software, 16 Middle St., Hallowell, Maine (Schecher and McAvoy, 1992, 1994).aMINEQL+ calculates the concentration of various chemical species at chemical equilibrium based on a database of equilibrium formation constants for most inorganic species encountered in natural waters under standard conditions. However, a number of errors present in the database may lead to propagated errors in chemical speciation calculations. The methods for choosing equilibrium constants from a standardised database and for performing the required modifications prior to insertion into the MINEQL+ database are described below.

Selecting Equilibrium Constants

In order to compute the concentrations of the different chemical species in the test medium at equilibrium, one must consider all the possible complexes that can be formed under the test conditions. For each complex, a formation constant must be defined in the MINEQL+ program database. The database published by the U.S. National Institute of Standards and Technology (NIST; Martell et al., 1998) is a near exhaustive collection of thermodynamic constants describing the complexation of metals with inorganic and organic ligands. This database is considered a "Standard Reference Data Base", and it succeeds earlier compilations of critical stability constants (Smith and Martell, 1974–1982). All equilibrium constants, enthalpy and entropy changes in the NIST database have been thoroughly scrutinised and objectively selected by the authors: bibliographic references are provided for each value. Values for constants that were in close agreement among different workers were given priority whenever possible. We have not critically compared the NIST (Martell et al., 1998) database with the MINEQL+ Ver.4.5 default database, but the NIST database is listed as a source document by Schecher (2001).

When selecting formation constants from the NIST database, priority should be given to those evaluated at the lowest ionic strength and at standard temperature (25°C). Consistency should also be observed in terms of the source of the data itself. Formation constants for a given metal–ligand system, one that is able to form numerous species in solution (ML, ML_2, ML_3, etc.), would ideally come from a single reliable source, where similar methodologies were used to determine each constant. For example, formation constants for the progressive hydrolysis of aluminium, $Al_3 + + _nOH \longrightarrow AlOH_2 +$, $Al(OH)_2 +$, $Al(OH)_3$ 0, $Al(OH)4$ 1–, would ideally be derived from the same data set and be in accordance with other data sets using similar methodologies.

Formation Constants

Most of the errors found in the MINEQL+ default database are related to a misunderstanding of the difference between an equilibrium constant and a formation constant. For example, let us consider the log K value given in MINEQL+ (v. 3.01) for the complex $CdHEDTA1-$ and compare this with the value given in the NIST reference database:

MINEQL+: $H^+ + Cd^{2+} + EDTA^{4-} \Leftrightarrow$
$\qquad\qquad CdHEDTA^{1-}$ $\qquad log\ K = 2.90$ $\hfill$ (1)

NIST: $\qquad H^+ + CdEDTA^{2-} \Leftrightarrow$
$\qquad\qquad CdHEDTA^{1-}$ $\qquad log\ K = 2.$ $\hfill$ (2)

Although at first glance these two reactions seem to be the same, they differ significantly. The reaction speci- fied in the NIST reference database represents the equilibrium constant of the reaction of a hydronium ion with the complex $CdEDTA^{2-}$. What is meant to be in the MINEQL+ database is a formation constant for the complete reaction of all the components taken separately as defined by the speciation model, in this case Cd^{2+}, $H+$ and $EDTA^{4-}$. Thus, all log K values to be used in the speciation model MINEQL+ must be expressed in terms of all components involved in the reaction. To express the log K value of the complex $CdHEDTA^{1-}$ properly, the NIST equilibrium constant must be combined with the equilibrium constant of the complex $CdEDTA^{2-}$ (1018.3, as expressed at infinite dilution; this will be examined later) for a resulting formation constant value of 1021.2, a difference of 18 orders of magnitude! This type of error occurs for complexation reactions that have more than two reactants.

Let us consider another example: CaHCitrate. The NIST database gives numerous constants for the reactions of Ca2+ with HCitrate^{2-} and H+ with Citrate3–, but only one of each at zero ionic strength (infinite dilution) and standard temperature. These equations are combined in the following manner:

$Ca^{2+} + HCitrate^{2-} \Leftrightarrow CaHCitrate$
$$\log K = 2.93, \Delta H = 0.78 \tag{3}$$

$H^+ + Citrate^{3-} \Leftrightarrow HCitrate^{2-}$
$$\log K = 6.40, \Delta H = -1.0 \tag{4}$$

$Ca^{2+} + H^+ + Citrate^{3-} \Leftrightarrow CaHCitrate$
$$\log K = 9.33, \Delta H = -0.22 \tag{5}$$

The resulting formation constant for the complex CaHCitrate is therefore 9.33 (equation (3) + equation (4)) with an enthalpy difference of –0.22, whereas the default log K value given by MINEQL+ is 3.02, a difference of six orders of magnitude in the strength of the complex, and the H is not defined by default. Although this type of error does not affect most organic and inorganic complexes dealt with in the MINEQL+ database, missing data can be another source of misunderstanding. The list of complexes in the MINEQL+ database is far from exhaustive, as some complexes are not defined in MINEQL+ (e.g., the complexation of trace metals by some amino acids). If these complexes are significant, in terms of speciation (as demonstrated in Twiss, 1996), then the predictions of the model will be erroneous. The user of any speciation model should compare all the default thermodynamic values, and the range of complexes formed, with a reliable source such as the NIST reference database (Martell et al., 1998).

Metal Hydrolysis

Another particularity of the MINEQL+ program concerns the convention for expressing the hydrolysis of ions. According to MINEQL+, the hydroxide ion does not exist alone but rather in the form of water. Thus, the formation of a hydroxo-complex must include the autoprotolysis of water, as shown in this example:

$$Zn^{2+} + OH^- \Leftrightarrow ZnOH^+ \qquad \log K = 5.0 \tag{6}$$

$$H_2O \Leftrightarrow OH^- + H^+ \qquad \log K = -14.0 \tag{7}$$

$$Zn^{2+} + H_2O \Leftrightarrow ZnOH^+ + H^+ \qquad \log K = -9.0 \tag{8}$$

Any constants to be used with MINEQL+ must be expressed in this form (8), where the formation of ZnOH$^+$ is derived from selecting Zn^{2+}, H$^+$ and H$_2$O as components in the program.

Solid Saturation

Formation of precipitates (defined as type V complexes) is indicated by MINEQL+ by a solid saturation index (SI). The system is undersaturated when the SI is negative and oversaturated when positive: a value of zero would indicate that equilibrium had been reached. The SI values for solids are shown in MINEQL+ in the log C column of the "output manager" section; this value should not be mistaken for a concentration. Although programs such as MINEQL+ can predict the formation of precipitates, some precipitates may take days, weeks or even longer to form. Ideally, chemical equilibrium simulations on the experimental media should not indicate any cases of oversaturation. In some cases (e.g., FeIII in many algal or higher plant culture media) this constraint may prove difficult to respect. In such cases, if empirical evidence indicates that the predicted precipitation of the dissolved solid is slow, then it can safely be ignored (empirical evidence in this case might be that all the metal remains in an operationallydefined dissolved state, as shown by filtration of the medium at the beginning and end of the experiment followed by total metal analysis of the respective filtrates).

Like formation constants, the solubility constants for complexes should be verified with a reliable source. A case in point is the mineral otavite ($CdCO_3$), the precipitation of which is predicted by MINEQL+ to occur at pH > 7 in systems that contain more than 0.85 M Cd and are open to the atmosphere. However, the solubility constant in the MINEQL+ (v. 3.01) default database is $10^{-13.74}$, much lower than the value of $10^{-12.00}$ quoted in the NIST database (Martell et al., 1998), and the value of $10^{-11.2}$ quoted by the CRC Handbook of Chemistry and Physics (CRC, 1994). In a careful literature review of the formation of otavite, Holm et al. (1996) reported values of solubility constants ranging from 10–13.74 to $10^{-11.28}$, whereas they independently established a value of 10–12.8. The constant introduced into the MINEQL+ default database was the lowest value ever published; this is an example of a poorly chosen constant. The selection of the proper solubility constant can therefore be as critical as the choice of a formation constant.

Expressing Formation Constants at Infinite Dilution

Speciation models such as MINEQL+ expect the formation constants to be expressed at infinite dilution. According to the ion concentrations specified by the user, the program will compute the ionic strength (I) and adjust the formation constant accordingly. Unless all the values are inserted at a uniform fixed ionic strength and the I correction function is not selected in the program, all values of formation constants in the thermodynamic database must be expressed at infinite dilution (I = 0). Several empirical approaches

(DebyeHuckel, Guntelberg and Davies) can be used to adjust an equilibrium constant as a function of the ionic strength (Morel and Hering, 1993; Stumm and Morgan, 1996). However, the Davies equation has the widest window of applicability, I from 0.5 to 0:

$$log\ \gamma_n = -A \cdot Z_i^2 \cdot [\ I^{1/2}/(1+I^{1/2}) - 0.2 \cdot I\ ] \tag{9}$$

where: γ_n= ionic activity coefficient for an n-charged species; Z_x = charge of the ion i; and A = temperature dependent constant (0.512 at 25°C; cf. Robinson and Stokes, 1965). Let us consider again the formation of the complex CdEDTA2–. To adjust the conditional equilibrium constant acquired from the NIST database from an ionic strength of 0.1 to I = 0, the following calculations are needed:

$$Cd^{2+} + EDTA^{4-}\ (\ CdEDTA^{2-} \tag{10}$$

$$^aK\ \ =\ \ \{CdEDTA^{2-}\}/\{Cd^{2+}\}\cdot\{EDTA^{4-}\}$$
$$\gamma_2/(\ \gamma_2 \cdot \gamma_4) \cdot\ ^cK \tag{11}$$

$$^aK = \gamma_2/(\ \gamma_2 \cdot \gamma_4) \cdot [CdEDTA^{2-}]/[Cd^{2+}]\cdot[EDTA^{4-}] \tag{12}$$

$$^aK = (1/10^{-1.8})\cdot 10^{16.5} = 10^{18.3};\ log\ K = 18.3 \tag{13}$$

where: ^aK = corrected formation constant at infinite dilution; ^cK = conditional constant, log ^cK = 16.5 at 25°C and μ = 0.1 (Martell et al., 1998)

Corrected formation constants can then be saved in a personal data file (personal.dat) in MINEQL+ by using the THERMSAVE function and be recalled later with the THERMREAD function (Schecher and McAvoy, 1994). Note that "personal" constants have to be recalled if a new component is added, since the default database must be reloaded for the changes to be effective.

Partial CO_2 Pressure

The default partial pressure of CO_2 (18.16) provided in the MINEQL+ program is equivalent to a one atmosphere partial pressure of CO_2, which does not correspond to the standard partial pressure of CO_2 in the earth's atmosphere. Calculations pertaining to the partial pressure of CO_2 are given in the instruction manual (Schecher and McAvoy, 1994). However, the correction must be made before a calculation is started. The value 21.66, corresponding to atmospheric pressure (Pco2 = 10-3.5), can also be stored by the THERMSAVE function. Omitting this correction could result in an overestimation of the concentration of carbonates and their metal complexes.

Temperature

All formation constants in the MINEQL+ database are, by default, valid at a temperature of 25°C. To insert a new constant that has been evaluated at a different temperature, the log Kvalue must be corrected using the enthalpy change. Likewise, to perform speciation calculations at temperatures other than 25°C, all the enthalpy changes specific to the reactions involved must be known. From the H value, the program will adjust the formation constant for any speci- fied temperature other than 25°C. If only some of the enthalpies are known then corrections of the log K values will also be partial and this will result in erroneous predictions; note that MINEQL+ will not alert the user when H values are missing.

The mathematical relationship expressing variations in an equilibrium constant as a function of temperature is given by the following equation:

$$\ln \frac{K_2}{K_1} = \int_{T_1}^{T_2} \frac{\Delta H_r^o}{RT^2} \delta T$$
(14)

Empirical approaches can be used to approximate this relationship such as the van't Hoff equation (Serkiz et al., 1996), which assumes the enthalpy change is constant over the temperature range T_1 to T_2:

$$\log K_{T2} = \log K_{T1} + H_r^o \cdot (T_2 - T_1) \cdot (0.00246)$$
(15)

where T is expressed in degrees Celsius.

Miscellaneous Considerations

Prior to an experiment, one must be aware of any possible photochemical reactions that could take place in the medium and any subsequent effect on speciation. Examples of photochemical reactions include the absorption of light by the Fe-EDTA complex, which leads to the oxidation of EDTA and the subsequent reduction of iron (Price et al., 1991), and the photo-oxidation of thiol-containing ligands such as cysteine and glutathione (Jocelyn, 1972; Scoffone et al., 1970). Another important aspect of speciation concerns the kinetics of the reactions. Speciation models such as MINEQL+ can predict concentrations at equilibrium, but they give no indication of the time necessary to achieve this equilibrium. "Sufficient" time allowed for a solution to reach equilibrium depends on the nature of the ligands used, on their tendency to form complexes with the major cations, Ca^{2+} and Mg^{2+}, and the respective concentrations of these "hardness" cations (cf. Hering and Morel, 1988).

ASSESSING THE BIOLOGICAL RESPONSE TO A TOXIC TRACE METAL

Having considered in some detail how exposure media should be prepared and manipulated in metal toxicity experiments, we now turn our attention briefly to the toxicological response itself and discuss the particularities of different biological "targets".

Phytoplankton: The choice of an endpoint to monitor metal toxicity to phytoplankton depends on which level the toxicity is to be studied. Classical metal toxicity studies are usually performed by following either the biomass after a specific exposure duration or the growth rate during the exponential growth phase; both parameters may be evaluated by the measurement of cell number, biomass, biovolume or chlorophyll content. Toxicity is expressed by the concentration of toxicant that inhibits growth or biomass by a given amount (often 50%) with respect to a control culture to which no toxicant was added, over a specific time period (e.g. 96-h IC_{50}).

The choice of response variable in toxicity tests using phytoplankton has been debated. Nyholm's (1985) analytical review summarises the controversy between the use of biomass and growth rate as end point criteria; supported by inter-laboratory tests and a theoretical standpoint, the use of growth rate is superior to biomass as an end-point criterion. Taken separately, these two end-point criteria could give different results. A good alternative solution, which combines these two parameters and the lag phase, is the use of the area under the growth curve along the incubation time necessary to reach the stationary growth phase (Nyholm, 1985, Parent and Campbell, 1994).

Solute transport into algal cells is a sensitive endpoint criterion since many metabolically important solutes (PO_4, NO_3) are affected by trace metal toxicity (cf. Petersen and Nyholm, 1993). Metal accumulation may also be used as a measure of interaction between the target organism and the trace metal. Generally, trace metal uptake is characteristically a bi-phasic process: a fast physical and/or chemical adsorption onto the cell surface, followed by a slower, facilitated transport into the cell (Bates et al., 1982; Schenck et al., 1988; Hudson and Morel, 1990). Total metal uptake can be measured directly on the algae and intracellular uptake determined after extracting the algae with strong ligands such as EDTA to remove surface-bound metal (Bates et al., 1982, Schenck et al., 1988).

One difficulty encountered when comparing published studies of algal metal uptake is the diversity of units used to express the results: metal per g of protein (e.g. Collard and Matagne, 1994), metal per kg of dry algal biomass (e.g. Sloof et al., 1995), metal per cell (e.g. Phinney and Bruland, 1994). For most studies, metal uptake results should be normalised per unit of cellular surface area and expressed as fluxes (e.g., mol $m^{-2} s^{-1}$) in order to facilitate inter-study comparisons. In some cases, however, the bioconcentration factor ([cell metal concentration] / [metal concentration in the exposure medium]) may be of diagnostic importance, in which case knowledge of the cell volume is required. If other expressions are used, then the necessary information should be presented to allow conversion of the data to a surface area basis. Surface areas can be calculated geometrically, on the basis of cell morphology and cell dimensions, either manually (microscopic examination) or with the aid of electronic particle counters, although the former is preferable (Hillebrand et al., 1999).

Rooted Plants

Measuring metal uptake by plant roots requires an understanding of the availability of the metal in the soil or sediment substrate. Typically, in metal bioaccumulation studies, metal uptake is related to some index of metal availability in the soil or sediment based on an elaborate extraction procedure. Many extraction procedures have been developed for metals (Soon and Abboud, 1993), but no one extraction procedure has been found to correlate adequately with plant metal accumulation over a wide range of soil properties. In addition, the chemical extractants alter the soil properties that control mineral solubility, pH, redox potential and organic matter solubility, which prohibits one from using the extraction data to estimate metal speciation in the original soil or sediment (McBride, 1994).

To really understand metal speciation in soils or sediments, and the mechanisms of metal uptake by plant roots, one has to consider in what forms the metal occurs in the soil solution or in sediment interstitial water, the medium from which nutrients are absorbed (Barber, 1995). Metals in soil solution, however, may occur in one of several oxidation states and in dissolved complexes with different inorganic or organic ligands. These considerations are important since the speciation of the metal will greatly influence its absorption by plant roots (McBride, 1994; Parker et al., 2001). The complex chemistry of soil solution and the dynamic processes of desorption, adsorption, precipitation reactions in soils have led researchers to develop artificial environments for studying metal accumulation or metal toxicity, using hydroponic solutions. The use of hydroponic solutions to study the accumulation of metal by plants has both advantages and disadvantages compared to the use of soils for such studies (Parker and Norvell, 1999). Several key issues should be considered, however, in order to maximise the advantages of using hydroponic solutions.

Media typically used to culture plants hydroponically may not be stable (e.g. Hoagland's solution; Hoagland and Arnon, 1950). Care must be taken to ensure that solubility constants are not exceeded by media components, especially for salts of iron or phosphate. Chemical equilibrium modelling programs such as MINEQL+ can be used to determine whether or not solubility limits are exceeded. The formation of precipitates in exposure solutions results in a situation where the precise dose and speciation of the metal being studied are no longer known, since a portion of the dissolved metal may precipitate (or become bound to precipitates) and become lost from solution. More complex, unstable media may be used to culture plants prior to exposure, but a simplified, stable media is recommended for exposure studies. Media lacking micronutrients, for example, are often stable and can be used for short-term exposures. Plant roots are known to excrete a range of organic compounds into the media immediately surrounding the roots (Cieslinski et al., 1997). These compounds modify solution pH and provide a source of ligands that have the potential to complex the metal being studied. These exudates may significantly alter speciation over time. This problem can be minimised by designing an exposure system with a relatively large {media volume: root mass} ratio (so that root exudates can be diluted by a large volume of solution), by designing experiments with short exposure duration (to reduce the amount of time exudation would have to alter solution speciation), or by adopting a recirculating or flowthough hydroponic system where exposure solutions are continually replenished. Cation-exchange columns can be used to estimate the proportion of free, cationic metal in solution and ensure that it is consistent over the duration of the exposure period (Fortin and Campbell, 1998).

By carefully considering these issues, it becomes possible to relate uptake of a metal by plant roots to the species of the metal that exist in the bulk solution surrounding the roots. It should be remembered, however, that the solution immediately surrounding plant roots may be somewhat different (pH, compounds secreted by plant roots), and that the magnitude of this difference is difficult to quantify and is not known at this time. Interesting results from hydroponic studies which carefully considered effects of metal speciation include the observation that the addition of chloride to the exposure solution resulted in enhanced accumulation of Cd by Swiss chard in relation to solution Cd^{2+} concentration (Smolders and McLaughlin, 1996a, 1996b). Increasing solution Cl- concentration resulted in the formation of $CdCl_n^{2-n}$ species, and the authors suggest that Cd accumulation was increased due to accumulation of these species or to enhanced diffusion of Cd from the bulk solution to the root surface. In a study on the effect of increasing solution $SO4^{2-}$ concentration, it was discovered that plant tissue Cd concentrations were unaffected by increasing solution $SO4^{2-}$ concentrations, even though solution Cd^{2+} dropped significantly, leading the authors to conclude that CdSO4 0 is taken up just as easily as Cd^{2+} (McLaughlin et al., 1998). In another study, Srivastava and Appenroth (1994) found that addition of EDTA to solutions containing Cd signifi- cantly reduced the Cd^{2+} activity, and also the accumulation of Cd by duckweed. However, the reduction in accumulation was not quite as much as predicted by the reduction in Cd^{2+} activity, and the authors attributed this result to uptake of Cd-EDTA species through breaks in the root endodermis or disassociation of CdEDTA during treatment. In studies where citrate was added to the exposure solution, it was discovered that accumulation of Cd by durum wheat roots did not decrease even though the free Cd^{2+} concentration dropped significantly as more Cd complexed with citrate (Berkelaar and Hale, 1998). Possible explanations for this observation included accumulation of Cdcitrate complexes, enhanced diffusion of Cd to uptake sites or a reduction in the free Ca^{2+} or Mg^{2+} concentrations in the external solution (these hardness cations may compete with Cd^{2+} for uptake sites).

Although hydroponic systems are ideal for quantifying metal accumulation they do not represent the same environment found in the soil solution of the rhizosphere. The rhizosphere has been defined as the zone of soil surrounding the root that is influenced by root processes (Darrah, 1993). The cylinder of soil that is influenced by such processes as nutrient diffusion and mass flow, root exudates, micro-organism activity and root uptake can vary from 1–2 mm to several cm from the root surface. Collection of rhizosphere soil, however, has proven to be a difficult task and is usually defined as the soil loosely adhering to the roots as they are being gently shaken (Marschner, 1995). The rhizosphere is a very dynamic zone in that water contents and ion concentrations will vary with the distance from the root due to diffusion gradients induced by uptake, making it difficult to quantify actual concentrations in the zone next to the root surface.

Roots also exude a number of compounds into the rhizosphere, among which the low molecular weight organic solutes are of most interest in terms of metal speciation. These include such compounds as sugars, organic acids, amino acids and phenolics (Marschner and Romheld, 1996). Low molecular weight organic acids have been studied in greater detail than the other exudates and have been shown to influence the speciation of Cd in solution (Mench et al., 1988) and in soil (Krishnamurti et al., 1996). However, due to difficulties in collecting rhizosphere soils and the time-consuming procedures for measuring low molecular weight organic acids (Szmigielska et al., 1996), little work has been done to determine free-metal ion concentrations in rhizosphere soil. Another concern in hydroponic and soil studies with regards to the free-metal ion concentration is that soils can buffer or continue to supply metals to soil solution whereas hydroponic solutions normally have a fixed total metal concentration, unless flowing solution cultures are used (Wild et al., 1979). Thus as the metal is depleted at the root surface either through uptake or complexation reactions, ion-exchange reactions will release more of the metal from the soil exchange sites into solution to reach the required equilibrium, hence the buffering effect. However, in hydroponic solutions, the continual release of root exudates can complex the metal thus reducing the free-metal ion concentration.

An example of the need to consider metal speciation in the rhizosphere is depicted in this example of Cd uptake modelling. Metal uptake models require an estimate of the initial metal ion concentration in soil solution (assumed to be the free-metal ion concentration) before the growth study starts as an input for the model simulations (Barber, 1995). Generally, metal concentrations are measured from soil solution using column displacement, centrifugation or immiscible displacement (Soon and Warren, 1993) but ion activities or chemical species are not calculated. Models of Cd uptake for corn have shown that Cd uptake was overpredicted and that the Cd solution concentration was the most influential parameter in varying simulated Cd uptake (Mullins et al., 1986). The over-prediction could be the result of not accounting for Cd speciation in the rhizosphere as affected by root exudates or other soluble organic matter in the rhizosphere, which would actually lower the free-metal ion concentration. The overprediction of Cd uptake in their study clearly suggests that knowledge of metal speciation is needed in order to accurately determine the concentration of the freemetal ion in solution in the rhizosphere that would be available for plant uptake in metal uptake modelling. An excellent review of chemical speciation and models for soil systems has been written by Loeppert et al. (1995).

Cultured Animal Cells

For animals, including humans, the question of primary interest is the precise evaluation of metal bioavailability following exposure to various metal species. To be biologically active in non-aquatic organisms, metals must normally be absorbed into the blood circulation before they are distributed to the various target organs where they exert their specific toxic effects. These toxic effects are generally well characterised but understanding metal interactions with biological membranes with respect to metal speciation and metal uptake by living cells remains a formidable challenge. In this context, experiments with animal and human cell cultures are very useful, since they allow the investigation of membrane processes and cellular mechanisms involved in metal uptake and cytotoxicity. However, a number of points should be considered with care. When studying metal passage through target epithelia (intestinal, pulmonary, gill) to investigate metal bioavailability, the first step is to evaluate the respective contributions of the paracellular and the transcellular pathways to transepithelial transport. Accordingly, the metal concentrations investigated should be below the threshold at which disruption of epithelium integrity occurs; epithelium permeability should be monitored with care. For epithelia that form tight junctions, evaluation of the transepithelial electrical resistance may be an adequate measure of epithelium integrity. The effects of metals on this parameter should be studied in relation to metal toxic effects.

Metal uptake and adsorption should clearly be distinguished. Generally, adsorption onto the outer membrane surface is defined as the labile fraction of metal that can be removed by washing with chelating agents such as EDTAor EGTA. Accordingly, "accumulation" should refer to the total amount of metal measured in cell samples (adsorption + uptake), "uptake" data correspond to metal that actually crosses the cell membrane (accumulation minus adsorption), whereas "adsorption" refers to the metal bound at the membrane external surface (accumulation minus uptake), that is the EDTA/EGTA-sensitive component of accumulation. However, we often neglect to consider that metal uptake is generally partially reversible, and that efflux may occur very rapidly, being significant within a few minutes. Indeed, in human intestinal cells Caco-2 Cd efflux was observed to proceed rapidly (Jumarie et al., 1997), this efflux not being exclusively related to desorption (Jumarie et al., 1999). This result demonstrates that metal taken up into the cells may be rapidly lost during the washing procedure, especially in the presence of EDTA/EGTA used as a chelator. Accordingly, the EDTA/EGTA extractable fraction should be interpreted with care since it is not necessarily related to adsorption exclusively, but may include an efflux component.

As indicated earlier, an exact interpretation of the toxicological data requires well-controlled metal speciation conditions and a complete knowledge of the chemical species present in the exposure medium. Ideally, this may be done using defined uptake media and chemical equilibrium programs to calculate metal speciation at equilibrium. The thermodynamic equilibrium constants required for equilibrium calculations are normally available for "standard conditions", e.g. 25°C, infinite dilution. However, these conditions are often far removed from those encountered in the exposure media used for cultured animals cells, e.g. 37°C and high ionic strengths, and the thermodynamic data needed to extrapolate from standard to experimental conditions are often unavailable. Under such conditions, the experimental determination of conditional constants for conditions representative of those used in the exposure experiments will be required (Jumarie et al., 2001).

Since most cultured animal cells require serum to grow and to undergo differentiation, and because serum may vary from one batch to another (e.g., containing different levels of protein likely to bind metal), a simple serum-free exposure medium should be established for which metal speciation can be calculated. For example, in Cd uptake studies performed on human intestinal cells, Caco-2, we used a serum-free minimal uptake medium containing NaCl, KCl, $CaCl_2$, $MgSO_4$ and D-glucose, plus HEPES as buffer (Jumarie et al., 1997). Note that the cells should be thoroughly washed before being exposed to the metal, in order to remove the remaining serum and to ensure that uptake / toxicity experiments are performed under well-controlled conditions.

When studying the impact of metal speciation on metal uptake, one often manipulates the exposure medium in order to vary metal speciation. However, for cultured animal cells one has to consider physiological parameters that may limit the range over which metal complexation can be varied. Cultured animal cells tolerate only small variations in pH, ionic strength, ionic gradient and temperature, and thus only limited modifications of the exposure media are allowed, which in turn may limit possible variations in metal speciation. One example that illustrates this situation well is the system involving exposure of human intestinal cells, Caco-2, to Cd and citrate. For these cells, as for all human cells, physiological values for the external osmolarity must be 295 ± 5 mOsm. Below this value, cells begin to shrink, whereas above 300 mOsm they rapidly undergo swelling. A number of regulatory mechanisms rapidly become operational when cell volume is modified, but these activities remain practically silent under normal conditions. The affinity of Cd for citrate is relatively low, and thus high amounts of citrate must be added to the exposure media to achieve significant concentrations of the Cd-citrate complex. This requirement leads to use of hyperosmotic media in experiments testing Cd-Cit uptake by Caco-2 cells. Accordingly, cells exposed to high Cd-citrate concentrations are also exposed to hyperosmotic media, which may, by itself, lead to modifications in metal uptake through active cell volume regulatory mechanisms. This example illustrates the need for adequate control experiments since, indeed, we have observed that hyperosmotic conditions tend to stimulate inorganic Cd uptake under similar speciation conditions.

CONCLUSIONS

The use of simplified, defined media for experiments on metal uptake, nutrition and toxicity experiments has become increasingly common over the past 20 years. Over this same time period, the computer programs used for the resolution of chemical equilibrium equations have migrated from the mainframe computer to the local server to the desktop personal computer. The combined use of these techniques has, however, been reserved for those possessing in-depth knowledge of both chemistry and biology. The present review should enable an investigator with basic biological, chemical and computing skills to design experiments that will yield unambiguous responses to the various hypotheses that are being tested in the field of trace metal interactions with living cells.

ACKNOWLEDGEMENTS

Funding for this review was provided from the Canadian Network of Toxicology Centres (PGCC, FD, BH) and a grant from the Natural Sciences and Engineering Research Council of Canada (MRT).

REFERENCES

1.   Barber, S.A. 1995. *Soil nutrient bioavailability: A mechanistic approach.* 2nd ed. John Wiley & Sons, New York, NY.

2.   Bates, S.S., Tessier, A., Campbell, P.G.C. and Buffle, J. 1982. Zinc adsorption and transport by *Chlamydomonas vari-abilis* and *Scenedesmus subspicatus* (Chlorophyceae) grown in semicontinuous culture. *J. Phycol.*, 18, 521–529.

3.   Berkelaar, E. and Hale, B. 1998. Influence of citrate on accumulation of cadmium by durum wheat. Poster presented at the SETAC 19th Annual Meeting, Charlotte, NC, USA, November 15–19, 1998.

4.   Block, M. and Pärt, P. 1986. Increased availability of cadmium to perfused rainbow trout (*Salmo gairdneri* Rich.) gills in the presence of the complexing agents diethyldithiocarbamate, ethyl xanthate and isopropyl xanthate. *Aquat. Toxicol.*, 8, 295–302.

5.   Bryan, G.W. 1971. Effects of heavy metals (other than mercury) on marine and estuarine organisms. *Proc. Roy. Soc. Lond.*, B 177, 389–410.

6.   Buffle, J. 1988. *Complexation reactions in aquatic systems: an analytical approach.* Ellis Horwood Ltd., John Wiley & Sons, Chichester.

7.   Campbell, P.G.C. 1995. Interactions between trace metals and aquatic organisms: a critique of the free-ion activity model. In: Tessier, A. and D.R. Turner (eds.) *Metal Speciation and Bioavailability in Aquatic Systems*, pp. 45–102. John Wiley & Sons, Chichester.

8.   Cieslinski, G., Van Rees, K.C., Szmigielska, A.M. and Huang, P.M. 1997. Low molecular weight organic acids released from roots of durum wheat and flax into sterile nutrient solutions. *J. Plant Nutr.*, 20, 753–764.

9.   Collard, J.M. and Matagne, R.F. 1994. Cd^{2+} resistance in wild type and mutant strains of*Chlamydomonas rein-hardtii*. *Environ. Exp. Bot.*, 34, 235–44.

10. CRC Handbook of Chemistry and Physics. 1994. D.R. Lide & H.P.R. Frederikse (75th ed.), p. B-207. Chemical Rubber Company, Inc., Boca Raton, FL.

11. Daly, H.R., Campbell, I.C. and Hart, B.T. 1990. Copper toxicity to *Paratya australiensis*: I. Influence of nitrilotri-acetic acid and glycine. *Envir. Toxicol. Chem.*, 9, 997–1005.

12. Darrah, P.R. 1993. The rhizosphere and plant nutrition: a quantitative approach. *Plant and Soil*,155/156, 1–20.

13. EPA. 1995. Errors found in speciation model. *Envir. Sci. Technol.*, 29, 66a.

14. Errécalde, O. and Campbell, P.G.C. 2000. Cadmium and zinc bioavailability to *Selenastrum capricornutum* (Chlorophyceae): accidental metal uptake and toxicity in the presence of citrate. *J. Phycol.*,36, 473–483

15. Errécalde, O., Seidl, M. and Campbell, P.G.C. 1998. Influence of a low molecular weight metabolite (citrate) on the toxicity of cadmium and zinc to the unicellular green alga *Selenastrum capricornutum*: an exception to the free-ion model. *Water Res.*, 32, 419–429.

16. Feng, Y. and Barratt, R.S. 1994. Lead and cadmium composition in indoor dust. *Sci. Tot. Environ.*,152, 261–267.

17. Ferguson, W.J., Braunschweiger, K.I., Braunschweiger, W.R., Smith, J.R., McCormick, J.J., Wasmann, C.C., Jarvis, N.P., Bell, D.H. and Good, N.E. 1980. Hydrogen ion buffers for biological research. *Anal. Biochem.*, 104, 300–310.

18. Fortin, C. and Campbell, P.G.C.. 1998. An ion-exchange technique for free-metal ion measurements (Cd^{2+}, Zn^{2+}): applications to complex aqueous media. *Int. J. Environ. Anal. Chem.*, 72, 173–194.

19. Good, N.E., Winget, G.D., Winter, W., Connolly, T.N., Izawa, S. and Singh, R.M.M. 1966. Hydrogen ion buffers for biological research. *Biochemistry*, 5, 467—477.

20. Guy, R.D. and Kean, A.R. 1980. Algae as a chemical speci-ation monitor-I. A comparison of algal growth and computer calculated speciation, *Water Res.*, 14, 891–899.

21. Hering, J.G. and Morel, F.M.M.. 1988. Kinetics of trace metal complexation: role of alkaline-earth metals. *Environ. Sci. Technol.*, 22, 1469–1478.

22. Hillebrand, H., Dürselen, C.-D., Kirschtel, D., Pollingher, U. and Zohary, T. 1999. Biovolume calculation for pelagic and benthic microalgae. *J. Phycol.*, 35, 403—424.

23. Hoagland, D.R. and Arnon, D.I. 1950. *The water-culture method for growing plants without soil.* Calif. Agric. Expt. Sta. Circular 347. Agricultural Productions, Univ. of California, Berkeley, CA.

24. Holm, P.E., Andersen, B.B.H. and Christensen, T.H. 1996. Cadmium solubility in aerobic soils. *Soil Sci. Soc. Amer. J.*, 60, 775–80.

25. Honeyman, B.D. and Santschi, P.H. 1988. Metals in aquatic systems. *Environ. Sci. Technol.*, 22, 862–871.

26. Hudson, R.J.M. and Morel, F.M.M. 1990. Iron transport in marine phytoplankton: kinetics of cellular and medium coordination reactions. *Limnol. Oceanogr.*, 35, 1002–1020.

27. Hume, D.N. 1973. Pitfalls in the determination of environmental trace metals. *Progr. Anal. Chem.*, 5, 3–16.

28. Jocelyn, P.C. 1972. Oxidation of thiols. In: *Biochemistry of the SH group*, Chapter 4, pp. 94–115. Academic Press, London.

29. Jumarie, C., Campbell, P.G.C., Berteloot, A., Houde, M. and Denizeau, F. 1997. Caco-2 cells used as an *in vitro* model to study cadmium accumulation in intestinal epithelial cells. *J. Membr. Biol.*, 158, 31–48.

30. Jumarie, C., Campbell, P.G.C., Houde, M. and Denizeau, F. 1999. Evidence for an intracellular barrier to cadmium transport through Caco-2 cell monolayers. *J. Cell Physiol.*, 180, 285–297.

31. Jumarie, C., Fortin, C., Houde, M., Campbell, P.G.C. and Denizeau, F. 2001. Cadmium uptake by Caco-2 cells: effects of Cd complexation by chloride, glutathione and phytochelatins. *Toxicol. Appl. Pharmacol.*, 170, 29–38.

32. Keller, M.D., Bellows, W.K. and Guillard, R.R.L.. 1988. Microwave treatment for sterilisation of phytoplankton culture media. *J. Exp. Mar. Biol. Ecol.*, 117, 279–283.

33. Krishnamurti, G.S.R., Huang, P.M. and Van Rees, K.C.J. 1996. Studies on soil rhizosphere: Speciation and availability of Cd. *Chem. Spec. Bioavail.*, 8, 23–28.

34. Kuwabara, J.S. 1985. Phosphorus-zinc interactive effects on growth by *Selenastrum capricornutum*(Chlorophyta). *Environ. Sci. Technol.*, 19, 417–421.

35. Lage, O.M., Vasconcelos, M.T.S.D., Soares, H.M.V.M., Osswald, J.M., Sansonetty, F., Parente, A.M. and Salema, R. 1996. Suitability of the pH

buffers 3-[JV-W-bis(hydroxymethyl)amino]-2-hydroxypropanesulfonic acid and V-2-hydroxyethylpiperazine-W'-2-ethanesul-fonic acid for *in vitro* copper toxicity studies. *Arch. Envir. Contam. Toxicol.*, 31, 199–205.

36. Langston, W.J. and Spence, S.K. 1995. Biological factors involved in metal concentrations observed in aquatic organisms. In: Tessier, A. and D.R. Turner (eds.) *Metal Speciation and Bioavailability in Aquatic Systems*, pp. 407–478. John Wiley & Sons, Chichester.

37. Lin, H. and Randall, D.J. 1990. The effect of varying water pH on the acidification of expired water in rainbow trout. *J. Exp. Biol.*, 149, 149–160.

38. Loeppert, R.H., Schwab, A.P. and Goldberg, S. 1995. *Chemical equilibrium and reaction models*. SSSA Special Publication No. 42. Soil Science Society of America, Madison, WI.

39. Marschner, H. 1995. *Mineral nutrition of higher plants*. 2nd ed. Academic Press, London.

40. Marschner, H. and Romheld, V. 1996. Root-induced changes in the availability of micronutrients in the rhizosphere. In: Y. Waisel, A. Eshel and U. Kafkafi (eds.) *Plant roots: The hidden half*. 2nd ed., pp. 557–579. Marcel Dekker Inc., New York, NY.

41. Martell, A.E., Smith, R.M. and Motekaitis, R.J.. 1998. *NIST Critical Stability Constants of Metal Complexes Database*. Version 5.0. U.S. Department of Commerce, Gaithersburg, MD.

42. McBride, M.B. 1994. *Environmental chemistry of soils*. Oxford University Press, New York, NY.

43. McLaughlin, M.J., Andrew, S.J. Smart, M.K. and Smolders, E. 1998. Effects of sulfate on cadmium uptake by Swiss chard: I. Effects of complexation and calcium competition in nutrient solutions. *Plant Soil*, 202, 211–216.

44. Mench, M., Morel, J.L., Guckert, A. and Guillet, B. 1988. Metal binding with root exudates of low molecular weight. *J. Soil Sci.*, 39, 521–527.

45. Morel, F.M.M. and Hering, J.G. 1993. *Principles and Applications of Aquatic Chemistry*, J. Wiley and Sons, >New York, NY.

46. Morel, F.M.M., Westall, J.C., Rueter, J.G. and Chaplick, J.P. 1975. *Description of the algal growth media Aquil and Fraquil*. Department of Civil Engineering, Massachusetts Institute of Technology, Technical note No.16, Cambridge, MA.

47. Morel, F.M.M., Rueter, J.G., Anderson D.M. and Guillard, R.R.L. 1979. Aquil: a chemically defined phytoplankton culture medium for trace

metal studies. *J. Phycol.*, 15, 135–141.

48. Mullins, G.L., Sommers, L.E. and Barber, S.A. 1986. Modelling the plant uptake of cadmium and zinc from soils treated with sewage sludge. *Soil Sci. Soc. Am. J.*, 50, 1245–1250.

49. Nichols, H.W. 1973. Growth media - freshwater. In: Stein, J.R. [ed.], *Handbook of Phycological Methods - Culture Methods and Growth Measurements*, pp. 7–24. Cambridge University Press, Cambridg.

50. Nriagu, J.O., Lawson, G., Wong, H.K.T. and Azcue, J.M. 1993. A protocol for minimising contamination in the analysis of trace metals in Great Lakes waters. *J. Great Lakes Res.*, 19, 175–182.

51. Nyholm, N. 1985. Response variable in algal growth inhibition tests - Biomass or growth rate? *Water Res.*, 19, 273–279.

52. Parent, L. and Campbell, P.G.C. 1994. Aluminum bioavail-ability to the green alga *Chlorella pyrenoidosa* in acidified synthetic soft water. *Envir. Toxicol. Chem.*, 13, 587–598.

53. Parker, D. R. and Norvell, W.A. 1999. Advances in solution culture methods for plant mineral nutrition research. *Adv. Agron.*, 65, 151–213.

54. Parker, D.R., Pedler, J.F., Ahnstrom, Z.A.S. and Resketo, M. 2001. Reevaluating the free-ion activity model of trace metal toxicity toward higher plants: experimental evidence with copper and zinc. *Environ. Toxicol. Chem.*, 20, 899–906.

55. Pärt, P., Svanberg, O. and Kiessling, A. 1985. The availability of cadmium to perfused rainbow trout gills in different water qualities. *Water Res.*, 19, 427–434.

56. Pärt, P. and Wikmark, G. 1984. The influence of some complexing agents (EDTA and citrate) on the uptake of cadmium in perfused rainbow trout gills. *Aquat. Toxicol.*, 5, 277–289.

57. Petersen, H.G. and Nyholm, N. 1993. Algal bioassays for metal toxicity identification. *Water Pollut. Res. J. Can.*, 28, 129–53.

58. Phinney, J.T. and Bruland, K.W. 1994. Uptake of lipophilic organic Cu, Cd and Pb complexes in the coastal diatom *Thalassiosira weissflogii*. *Environ. Sci. Technol.*, 28, 1781–90.

59. Playle, R.C. and Wood, C.M. 1989. Water chemistry changes in the gill micro-environment of rainbow trout: experimental observations and theory. *J. Comp. Physiol.*, 159 B, 527–537.

60. Poldoski, J.E. 1979. Cadmium bio-accumulation assays-their relationship to various ionic equilibria in Lake Superior water. *Envir. Sci. Technol.*, 13, 701–706.

61. Price, N.M., Harrison, G.I., Hering, J.G., Hudson, R.J., Nirel, P.M.V., Palenik, B. and Morel, F.M.M. 1991. Preparation and chemistry of the artificial algal culture medium Aquil. *Biol. Oceanogr.*, 6, 443–461.

62. Randall, D., Lin, H. and Wright, P.A. 1991. Gill water flow and the chemistry of the boundary layer.*Physiol. Zool.*, 64, 26–38.

63. Robinson, R.A. and Stokes, R.H. 1965. *Electrolyte solutions*, Butterworths, London.

64. Schecher, W.D. 2001. *Thermodynamic data used in MINEQL+ version 4.5*. Environmental Research Software, Hallowell, ME.

65. Schecher, W.D. and McAvoy, D.C. 1992. MINEQL+: a software environment for chemical equilibrium modelling. *Comput. Envir. Urban Systems* 16, 65–76.

66. Schecher, W.D. and McAvoy, D.C. 1994. *MINEQL+: A Chemical Equilibrium Program for Personal Computers* (Version 3.01). Environmental Research Software Hallowell, ME.

67. Schenck, R.C., Tessier, A. and Campbell, P.G.C. 1988. The effect of pH on iron and manganese uptake by a green alga. *Limnol. Oceanogr.*, 33, 538–550.

68. Scoffone, E., Jori, G. and Galiazzo, G. 1970. Selective photo-oxidation of amino acids in proteins.*Biochem. Soc. Symp.*, 31, 163–170.

69. Serkiz, S.M., Allison, J.D., Perdue, E.M., Allen, H.E. and Brown, D.S. 1996. Correcting errors in the thermo-dynamic database for the equilibrium speciation mode MINTEQA2. *Water Res.*, 30, 1930–1933.

70. Simkiss, K. and Taylor, M.L. 1995. Transport of metal across membranes. In: Tessier, A. and D.R. Turner (eds.) *Metal Speciation and Bioavailability in Aquatic Systems.*, pp. 1–44. J. Wiley and Sons, Chichester.

71. Sloof, J.E., Viragh, A. and Van Der Veer, B. 1995. Kinetics of cadmium uptake by green algae. *Water Air Soil Pollut.*, 83, 105–122.

72. Smith, R.M. and Martell, A.E. 1974–1982. *Critical Stability Constants*, vol. 1–5. Plenum Press, New York, NY.

73. Smolders, E. and McLaughlin, M.J. 1996a. Chloride increases cadmium uptake in Swiss chard in a resin-buffered nutrient solution. *Soil Sci. Soc. Am. J.*, 60, 1443–1447.

74. Smolders, E. and McLaughlin, M.J. 1996b. Effect of Cl on Cd uptake by Swiss chard in nutrient solutions. *Plant and Soil*, 179, 57–64.

75. Soares, H.M.V.M., Conde, P.C.F.L., Almeida, A.A.N. and Vasconcelos, M.T.S.D. 1999. Evaluation of amino-sulfonic acid pH buffers with

a morpholinic ring for cadmium and lead speciation studies by electroanalytic techniques. *Anal. Chim. Acta*, 394, 325–335.

76. Soon, Y.K. and Abboud, S. 1993. Cadmium, chromium, lead, and nickel. In: M.R. Carter (ed.) *Soil sampling and methods of analysis*, pp. 101–108. Lewis Publishers, Boca Raton, FL.

77. Soon, Y.K. and Warren, C.J. 1993. Soil solution. In: M.R. Carter (ed.) *Soil sampling and methods of analysis*, pp. 147–159. Lewis Publishers, Boca Raton, FL.

78. Srivastava, A. and Appenroth, K.J. 1994. Interaction of EDTA and iron on the accumulation of Cd^{2+} in Duckweeds *(Lemnaceae)*. *J. Plant Physiol.*, 146, 173–176.

79. Stumm, W. and Morgan, J.J. 1996. *Aquatic Chemistry -Chemical Equilibria and Rates in Natural Waters*. Third Edition. J. Wiley and Sons Ltd., New York, NY.

80. Sunda, W.G. 1991. Trace metal interactions with marine phytoplankton. *Biol. Oceanogr.*, 6, 411–442.

81. Sunda, W.G. and Huntsman, S.A. 1983. Effect of competitive interactions between Mn and Cu on cellular Mn and growth in estuarine and oceanic species of the diatom *Thalassiosira*. *Limnol. Oceanogr.*, 28, 924–934.

82. Sunda, W.G. and Lewis, J.A.M. 1978. Effect of complexation by natural organic ligands on the toxicity of copper to a unicellular alga *Monochrysis lutheri*. *Limnol. Oceanogr.*, 23, 870–876.

83. Szmigielska, A.M., Van Rees, K.C.J., Cieslinski, G., Huang, P.M. 1996. Low molecular weight dicarboxylic acids in rhizosphere soil of durum wheat. *J. Agric. Food Chem.*, 44, 1036–1040.

84. Taylor, G.J. and Foy, C.D. 1985. Mechanisms of Al tolerance in *Triticum aestivum* (wheat). IV. The role of ammonium and nitrate nutrition. *Can. J. Bot.*, 63, 2181–2186.

85. Tessier, A. and Turner, D.R. (eds.). 1995. *Metal Speciation and Bioavailability in Aquatic Systems*. John Wiley & Sons, Chichester, UK.

86. Twiss, M.R. 1996. The importance of chemical speciation: from the bulk solution to the cell surface. *J. Phycol.*, 32, 885–886.

87. Twiss, M.R. and Nalewajko, C. 1992. Influence of phosphorus nutrition on copper toxicity to three strains of *Scenedesmus acutus* (Chlorophyceae). *J. Phycol.*, 28, 291–298.

88. Vasconcelos, M.T.S.D., Azenha, M.A.G.O. and Lage, O.M. 1996. Electrochemical evidence of surfactant activity of the HEPES pH buffer

which may have implications on trace metal availability to cultures*in vitro. Anal. Biochem.*, 241, 248–253.

89.   Vasconcelos, M.T.S.D. and Almeida, A.A.N. 2000. Influence of zwitterionic pH buffers on the bioavailability and toxicity of copper to the alga *Amphidinium carterae. Environ. Toxicol. Chem.*, 19, 2452–2550.

90.   Wilhelm, S.W. 1995. Ecology of iron-limited cyanobacteria: a review of physiological responses and implications for aquatic systems. *Aquat. Microb. Ecol.*, 9, 295–303.

91.   Wild, A., Woodhouse, P.J. and Hopper, J.J. 1979. A comparison between the uptake of potassium by plants from solutions of constant potassium concentration and during depletion. *J. Expt. Bot.*, 30, 697–704.

92.   Xue, H.B., Stumm, W. and Sigg, L. 1988. The binding of heavy metals to algal surfaces. *Water Res.*,22, 917–26.

Chapter 3

COMPUTER SIMULATION OF DYNAMICAL ANOMALIES IN STRETCHED WATER

P. A. Netz[I]; F. Starr[II]; M. C. Barbosa[III]; H. Eugene Stanley[IV]

[I]Departamento de Química, ULBRA, Canoas RS, Brasil and Departamento de Química, Unilasalle, Canoas RS, Brasil

[II]Center for Theoretical and Computational Materials Science and Polymers Division, National Institute of Standards and Technology, Gaithersburg, Maryland 20899, USA

[III]Departamento de Física, UFRGS, Porto Alegre RS, Brasil

[IV]Center of Polymer Studies - Boston University Boston, MA 02215, USA

ABSTRACT

In this work, we describe how the anomalous diffusivity is related to the structural anomalies. For this purpose, we study how the thermodynamics and the dynamics of low-temperature water are affected by the decrease of the density.

INTRODUCTION

Water is one of the most important substances in nature, and its remarkably complex behavior is puzzling, specially when we consider the simplicity of the chemical structure of water's molecule [1-4]. Water behaves anomously in several ways[5]: it expands on freezing and therefore has a negative slope in the solid-liquid equilibrium line in the P-T diagram. Under atmospheric pressure, water has a density maximum at 4°C, a minimum in the isothermal compressibility at 46°C, and a minimum in the isobaric heat capacity at 35°C. It has usually high melting, boiling and critical points, among several other anomalies[6, 7].

It is known that these anomalies, as well as the main properties of water, including the properties that make water the essential fluid in biological systems, are linked to the microscopic structure of water and the peculiar intermolecular interactions. Nonetheless, it remains astonishing, how does this complex behavior emerge from a simple structure. The reproduction of the anomalous behavior and the description of these complexities is a challenge not only to the computer simulations, but also to the theoreticians that are developing theories for the mechanisms that govern complex systems. And in this aspect, computer simulations have been sucessfull in reproducing a wide range of properties of water and several anomalies, starting with very simple models.

Particularly, it is already known that there are three kinds of anomalies[8]: thermodynamic, dynamical and structural. The region in the phase diagram where the first kind of anomaly occurs is entirely inside the domain of the second, which in its turn within the region of the third. Inside the region of structural anomalies, the orientational and translational order show inter-dependence[8]. A key point is, therefore, to discover how does the structure affect the mobility.

The complex behavior of water can be approached both from the thermodynamic as well as from the microscopic level. Let us begin by focusing in the thermodynamical aspects. There are three possible explanations for the anomalous increase of the thermodynamic response functions (such as isothermal compressibility, isobaric heat capacity) on cooling, namely, the stability limit conjecture [9-11], the critical point hypothesis [12-15] and the singularity-free hypothesis [16-18]. According to the stability-limit conjecture, the increase in the thermodynamic response functions on cooling can be explained supposing that the curves are close to the spinodal, the line separating the region in the phase diagram where the liquid phase is metastable from the region where it is unstable. The pressure in the liquid spinodal in water's phase diagram, according to this conjecture, decreases on cooling, but attains a minimum value (at negative pressures). The spinodal reenters the positive pressures region of the phase diagram with further cooling, being in this way called a reentrant spinodal. The critical point hypothesis explains the anomalous increase in the response functions as being due to the proximity of a second critical point, which is located probably at - 85°C and 230 MPa[19], at the end of a line of first-order phase transition separating two liquid phases of different density. According to the singularity-free hypothesis, there is actually no divergence in the values of the response functions, and no need to postulate a proximity of spinodal or critical point. These functions remain finite and attain maximum values. The two last hypotheses are consistent with a non-reentrant spinodal.

The origin of these anomalies are in the peculiar microscopic structure of liquid water. Water can be regarded as a transient gel[16, 20], a highly associated liquid with strongly directional hydrogen bonds. These bonds induce a local order, however not strong enough, so water maintains a long-range disorder typical of liquids. The local structure and therefore also the anomalies are enhanced when water is subject to stretching, as well as strong cooling[21, 22]. In this sense, some very important insights in water's behavior can be obtained by studying supercooled water under negative pressures (stretching). Fluids under negative pressures are relevant not only from the academic point-of-view (it is not merely an academic curiosity), but also play an important role on realistic system, such as in the transport of water in plants[23]. Experimental results [24-26] as well as computer simulations [27-35] show that, starting with atmospheric conditions of pressure, the gradual increase of pressure increases the number of defects such as multiple H-bonds and interstitial water molecules, disrupting the tetrahedral local structure and leading to a weakening of the H-bonds and therefore to an increase in the mobility [32-39]. Applying very high pressures, however, leads to steric effects which lower the mobility, and as a result the diffusion constant attains a maximum at a given density r_{max}. Above r_{max}, the diffusion of water is controlled by hindrance, with the hydrogen bonds playing a secondary role. We show that a complementary behavior is found if water is submitted to a gradual decrease in pressure, from atmospheric to negative pressures, with a minimum of diffusivity at a given density r_{min}, as discussed in details further in this paper.

There are many different models used in the computer simulation, and the differences between them were recently reviewed[40, 41] None of them can reproduce exactly all of water properties and anomalies. Even though, the overall thermodynamic picture that emerges from different computer simulation models is roughly the same, but the dynamic properties are slightly more model-dependent. From the large diversity of potentials, the potentials with three interaction centers are computationally faster. Among them the SPC/E[42] is particularly accurate in the description of thermodynamic and dynamical properties of water and is largely used in pure water simulations as well as in biological systems. It is also known that the SPC/E water model can reproduce the maximum in diffusivity under pressure as well as the power-law behavior of dynamical properties on cooling and therefore it seems to be a good choice for analyzing the dynamics of stretched water.

In this work, we will describe how the anomalous diffusivity is related to the structural anomalies. For this purpose, we study how the thermodynamics and the dynamics of low-temperature water are affected by the decrease of the density.

METHODS

We performed an extensive set of molecular dynamics simulations using 216 SPC/E [42] water molecules. The Newton equations of motion were integrated using SHAKE [43, 44] with time steps of 1.0 fs for T > 210 K and 2.0 fs for T = 210 K. The range of temperatures and densities covered was 210 K < T < 280 K and 0.825 g/cm³ < r < 1.30 g/cm³. Many state points in this range have negative pressure, and at low densities they are either a metastable stretched liquid, or a phase separated liquid-gas mixture. All simulations were carried out in the canonical ensemble (NVT), in a cubic simulation box using periodic boundary conditions. The rescaling of the velocities was made using the Berendsen thermostat [45], the electrostatic interactions were calculated using reaction field [46] with cut-off radius of 0.79 nm.

The diffusion coefficient D was calculated from the asymptotic slope of the mean square displacement plotted versus time:

$$D = \frac{1}{6} \frac{d\langle r^2(t) \rangle}{dt}$$

(1)

The orientational relaxation was analyzed using the rotational autocorrelation functions [43]:

$$C(\mathbf{e}) = \langle \mathbf{e}(t) \cdot \mathbf{e}(0) \rangle$$

(2)

The vector $\mathbf{e}$ is a chosen unity vector describing the orientation of the dipole moment. Other possible choices of unity vectors such as the O-H bond direction and the vector perpendicular to the plane of the molecule give similar results[48]. The orientational relaxation time was calculated from these functions using an biexponential decay fit function. [47]

$$C = a_0 \exp(-bt^2/2) + a_I \exp(-t/\tau^I) + a_{II} \exp(-t/\tau^{II})$$

(3)

The first term correspond to a fast librational motion, and the two relaxation times take into account the possibility of a fast and a slow reorientation process. However, in the most of the cases taking only one exponential yielded a reasonable result, and thus only one relaxation time was determined.

In order to understand the effect of the structure on the dynamics we also carried out a detailed analysis of the local structure of water [49] computing the distribution of the number and angle of H-bonds (O-H ... O) and distribution of the number and angle of first neighbor molecules, analyzing the angle O ... O ... O. The number of first neighbors is calculated by computing all molecules whose distance from a given molecule is less than 0.32 nm. This distance represents the first minimum in the oxygen-oxygen radial distribution function g_{OO} (r) at moderate and low densities. In order to have a better comparison, the same O-O distance criterion is also applied for all state points, irrespective the density.

The distribution of the number of hydrogen bonds is computed in a similar way. A hydrogen bond between two water molecules is counted if an O-O pair within 0.32 nm and a H...O with a bound distance less than 0.25 nm are found[50]. This criterium is similar to the geometrical definition of the hydrogen bonds [51, 52] but modified in the sense that no restriction about hydrogen bond angle was applied.

RESULTS

Thermodynamic Properties

Shows the thermodynamic properties obtained in our MD simulations. A crucial point in the distinction between the different thermodynamic scenarios, the stability-limit conjecture the critical point and the singularity-free hypotheses, is the shape of the spinodal. Computer simulations help us to locate the spinodal. We fit the P-risotherms with a fifth-order polynomial and locate the minima in each isotherm, where the isothermal compressibility is zero. These minima define the stability limit beyond which no equilibrium exists and therefore it locates the spinodal $P_{sp}(T)$. We found a non-reentrant spinodal[53], qualitatively similar to that found by Harrington and coworkers[31]. The same result was also found by Yamada and coworkers [54-56] using the TIP5P[57] potential. This behavior can support both the critical point hypothesis and the singularity-free scenario.

The density at the spinodal $r_{sp}(T)$ was calculated as well and is located in a narrow range $0.853 \leq r_{sp} \leq 0.874$. For our results that means that our simulated points whose density is less than 0.875 either should be ruled out or care must be taken in order to check the evidence of cavitation.

Table 1: Thermodynamic properties: temperature, density, potential energy and pressure. The errorbars in the energy are about 0.2 and in the pressure, about 15-20

T (K)	ρ (g/cm³)	U (kJ/mol)	P (MPa)
280	0.850	- 47.01	- 239
	0.875	- 47.32	- 230
	0.900	- 47.75	- 204
	0.925	- 47.90	- 172
	1.125	- 48.86	320
260	0.850	- 48.53	- 261
	0.875	- 48.88	- 257
	0.900	- 49.23	- 231
	0.925	- 49.48	- 191
	1.125	- 50.01	289
250	0.850	- 49.28	- 273
	0.875	- 49.76	- 271
	0.900	- 50.03	- 244
	0.950	- 50.40	- 148
240	0.850	- 50.17	- 263
	0.875	- 50.51	- 282
	0.900	- 50.98	- 258
	0.925	- 51.30	- 213
	0.950	- 51.26	- 155
	0.975	- 51.32	- 102
	1.000	- 51.38	- 36
	1.075	- 51.31	134
	1.125	- 51.27	271
	1.250	- 51.24	758
	1.300	- 51.26	1042
230	0.850	- 51.06	- 278
	0.875	- 51.28	- 300
	0.900	- 51.80	- 272
	0.925	- 52.01	- 212
	1.125	- 51.92	230
220	0.900	- 52.62	- 278
	0.925	- 53.14	- 221
210	0.850	- 53.15	- 344
	0.875	- 52.92	- 348
	0.925	- 53.55	- 225

Dynamic Properties

The translational diffusion coefficient (D) and the orientational relaxation time (τ) of SPC/E water were analyzed. Table 2 shows the dynamic properties of the simulated state points. Fig. 1 shows D along isotherms. For $T \le 260$ K, D has a minimum at about r » 0.9 (g/cm³), which becomes more pronounced at lower temperatures. The rotational diffusion (orientational relaxation time) displays a very similar trend, as shown in Fig. 2. The behavior of τ along isotherms is in fact complementar to the behavior of D. The relaxation time τ initially increases, with decreasing density, passes through a maximum and decreases with further stretching. the maximum in τ can be found in the same region as the minimum in D. The product $\tau \times D$ (not shown) is remarkably constant irrespective the temperature or density[58, 50].

Table 2: Temperature, density, diffusion coefficient and orientational relaxation time for the dipole vector

T (K)	ρ (g/cm^3)	D (10^{-5} cm^2/s)	τ (ps)
280	0.850	1.359	9.16
	0.875	1.281	9.19
	0.900	1.261	9.46
	0.925	1.234	9.12
	1.125	1.158	6.59
260	0.850	0.634	23.91
	0.875	0.531	23.47
	0.900	0.527	22.16
	0.925	0.500	26.93
	1.125	0.635	12.35
250	0.850	0.346	40.00
	0.875	0.298	47.69
	0.900	0.295	43.89
	0.950	0.281	38.73
240	0.875	0.130	95.89
	0.900	0.105	118.7
	0.925	0.122	103.8
	0.950	0.149	75.37
	0.975	0.178	66.06
	1.000	0.210	56.63
	1.075	0.293	34.71
	1.125	0.291	28.85
	1.250	0.210	27.73
	1.300	0.153	35.29
230	0.850	0.0674	216.8
	0.875	0.0601	239.8
	0.900	0.0435	314.3
	0.925	0.048	238.8
	1.125	0.178	48.61
220	0.900	0.0114	623.9
	0.925	0.00612	1208
210	0.850	0.0019	9229
	0.875	0.00373	3582
	0.925	0.00303	3441

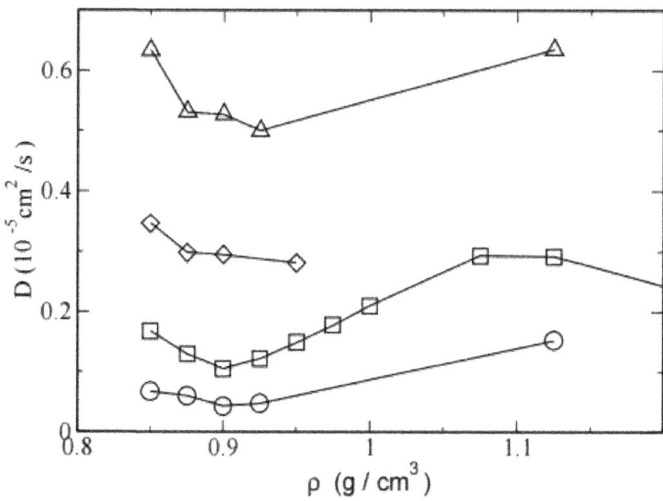

Figure 1: Diffusion Coefficient versus Density for temperatures T = 230K (circles), T = 240K (squares), T = 250K (diamonds) and T = 260K (triangles).

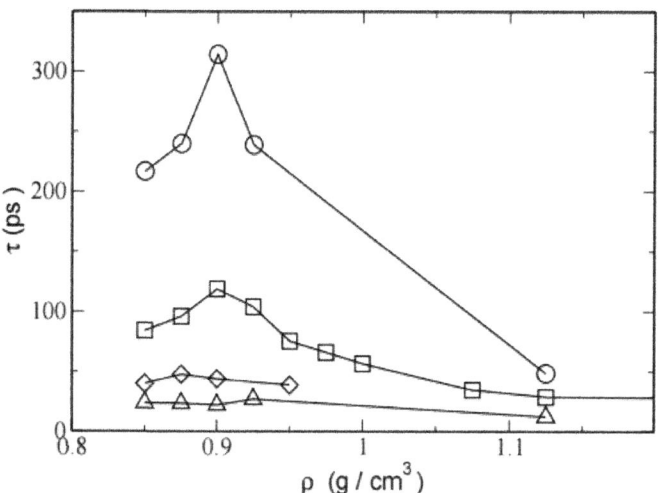

Figure 2: Orientational Relaxation Time versus Density for temperatures T = 230K (circles), T = 240K (squares), T = 250K (diamonds) and T = 260K (triangles).

These results show that the stretching has influence both in the translational as well as in the rotational diffusion. The complementar behavior of translational and rotational diffusion should be interpreted not in terms of hydrodynamic arguments[59], but rather as a clue pointing out that both the translational diffusion as well as the orientational process involve microscopic rearrangements, i.e. both motions are correlated by a common mechanism. In order to better understand this mechanism, the detailed microscopic structure of water was also investigated.

Structural Properties

One important aspect about molecular dynamics simulations is the very detailed microscopic information that these simulations can give. For instance, it is possible to look at the neighbor structure of each water molecule at each simulation step, obtaining a statistical picture of the local structure. Several aspects can be investigated, such as the distribution of the number of neighbors and number of H-bonds and also the angle distribution of neighbor molecules (O ... O ... O) angle and angle distribution of H-bonds.

Table 3: Results of the simulations. Distribution of the number of H bonds and neighbors

T	ρ	n_{HB}				n_{neighb}			
		≤ 2	3	4	≥ 5	≤ 2	3	4	≥ 5
280	0.850	6.6	28.1	58.2	7.1	3.9	22.4	56.5	17.6
	0.875	6.3	27.3	58.3	8.2	3.6	21.2	56.2	19.0
	0.900	5.1	24.6	62.2	8.0	2.7	17.8	58.0	19.4
	0.925	5.1	24.2	61.3	9.4	2.6	17.3	56.4	20.7
	1.125	3.4	19.6	56.4	20.7	1.3	10.5	44.1	44.0
260	0.850	4.8	24.0	65.1	6.1	3.2	18.9	62.4	14.5
	0.875	3.8	22.6	67.2	6.4	2.1	17.7	63.8	16.4
	0.900	3.9	20.4	68.9	6.8	2.1	15.4	64.0	18.6
	0.925	3.1	18.9	70.2	11.1	1.4	13.5	64.7	20.4
	1.125	2.5	17.6	60.3	19.6	0.9	9.5	47.4	42.3
250	0.850	3.5	21.5	69.9	5.2	2.0	17.3	66.5	14.2
	0.875	3.1	19.7	72.2	6.6	1.8	15.2	68.2	14.6
	0.900	2.8	18.2	72.8	6.3	1.5	13.5	68.4	16.4
	0.950	2.4	17.3	72.1	8.2	1.1	i 11.8	64.5	22.6
240	0.850	2.9	18.1	74.4	4.5	1.6	14.4	70.5	15.5
	0.875	2.4	17.2	75.8	4.6	1.3	13.5	72.5	12.7
	0.900	2.1	15.2	78.2	4.6	1.1	11.5	73.9	13.5
	0.925	1.8	14.3	78.4	5.6	0.8	10.4	72.9	16.0
	0.950	1.9	14.3	76.8	9.6	0.7	9.7	69.1	20.5
	0.975	2.5	17.2	73.0	7.2	1.5	14.1	70.3	14.1
	1.000	2.3	15.9	73.0	8.8	1.3	12.5	69.7	16.5
	1.075	1.5	12.9	70.6	15.0	0.3	5.2	48.8	45.8
	1.125	1.8	14.8	66.2	17.1	0.5	8.1	51.4	40.0
	1.250	1.2	11.7	56.7	30.5	0.1	3.2	27.9	68.7
	1.300	0.9	9.3	53.8	36.0	0.1	1.8	19.1	79.1
230	0.850	2.3	15.3	78.8	3.7	1.3	12.4	74.9	11.4
	0.875	1.8	14.4	80.1	3.8	0.9	11.2	77.1	10.9
	0.900	1.4	12.7	82.3	3.6	0.6	9.8	78.5	11.1
	0.925	1.4	11.7	82.1	4.8	0.5	8.5	77.0	14.0
	1.125	1.6	13.6	69.0	15.0	4.7	7.4	54.7	37.5
220	0.900	1.1	9.6	86.3	3.0	0.4	7.3	82.6	14.2
	0.925	0.9	7.9	88.2	3.1	0.3	5.7	83.6	10.5
210	0.850	1.1	8.5	89.0	1.5	0.5	6.9	87.2	5.5
	0.875	1.0	9.1	88.1	1.9	0.3	7.3	85.5	6.9
	0.925	0.6	4.4	92.6	2.4	0.0	1.5	73.7	24.8

Table 3 shows the distribution of the number of neighbors and H-bonds for the simulated state points. Fig. 3shows the H-bond and O ... O ... O angle distributions for the temperature $T = 240$ K, at several densities, whereas the same properties, analysed fixing the density at r = 0.90 g/cm^3 and at r = 1.125 g/cm^3 at several temperatures were shown in Figs. 4 and 5, respectively.

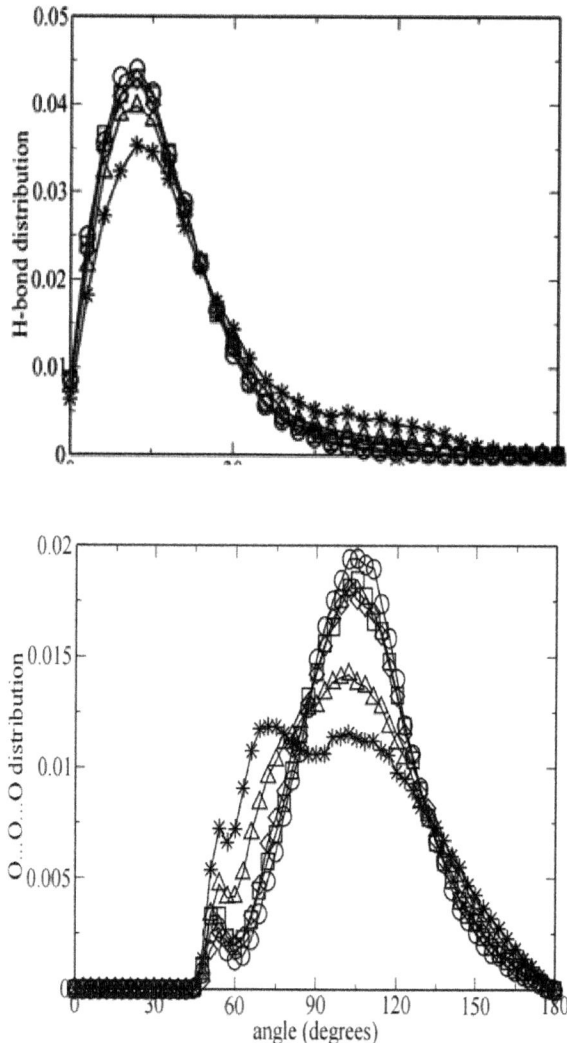

Figure 3: (a) Hydrogen bond angle distribution and (b) neighbors O · · · O · · · O angle distribution for several densities at the temperature T = 240 K. Shown are the results for ρ = 0.900 g cm−3 (circles), 0.950 (squares), 1.000 (diamonds), 1.125 (triangles) and 1.250 g cm−3 (stars). F

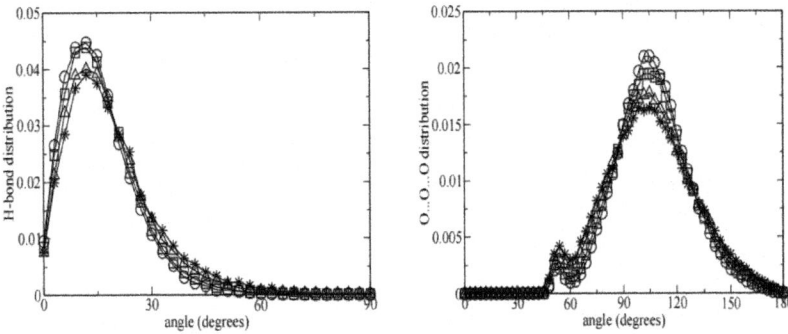

Figure 4: The same as Fig. 3, but for several temperatures at fixed $\rho = 0.900$ g cm^{-3}. Shown are the results for T = 230 K (circles), T = 240 K (squares), T = 260 K (triangles) and T = 280 K (stars)

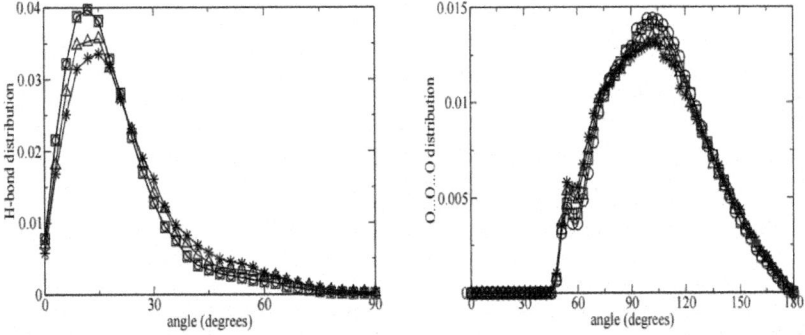

Figure 5: The same as Fig. 4, but for $\rho = 1.125$ g cm^{-3}. The symbols are the same as in Fig. 4.

The detailed analysis of the local structure of water reveals the enhancement of the tetrahedrality at low temperatures and ice-like densities. This can be detected from the angle distributions shown in these figures. Lowering the density, the O ... O ... O angle distribution becomes more sharply peaked around the tetrahedral angle, indicating an enhancement of the ice-like structure. the peak around 60°, on the other side, is related to a fifth neighbor in the coordination shell and the presence of this fifth neighbor increases with increasing density, as shown in Fig. 3b and Table 3. At very high densities, a peak appears around 80°, related to the appearance of a 6th neighbor[50]. The O ... O ... O angle distribution, at a fixed density, seems to depend only weakly on temperature, as shown in Figs. 4b and 5b.

The H-bond angle distribution, indicated in Fig. 3a seems not to depend on density, as far as this density is not too high. Fixing the density and analyzing the effect of temperature (as in Figs. 4a and 5a) we see that the peaks become broad with increasing T, indicating only a thermal fluctuation. Only at very high densities does the H bond become distorted.

The increase of the number of molecules with higher coordination numbers at densities ranging between r_{min} and r_{max} is responsible for the increase on the diffusion coefficient with increasing density. The number of H-bonds, however, does not alter significantly, meaning that these imperfections arise from the inclusion of extra molecules in the coordination shell which either do not for H-bonds or share and H-bond with another molecule. This shared bonds are weakened and the molecule can connect to another molecule by means of a small rotation, keeping the remaining H-bonds. The net motion of water structure can be described as a slow non-oscillatory or quasi-oscillatory translational and rotational displacements[60]. This peculiar mechanism can explain the coupling between translational and rotational diffusion, without the need of invoking hydrodynamical arguments.

CONCLUSIONS

In conclusion, we propose that the anomalous dynamic behavior of supercooled water can be explain in terms of its structural properties: the number and the distribution of hydrogen bonds and and the number and angular distribution of the O-O-O neighbors. We calculate these quantities for a wide range of densities and temperatures. We observe that the number of molecules with four neighbors has a maximum at the density r_{min} where the translational diffusion has a minimum and the rotational diffusion is maximum. For r_{max}, the density of highest translational diffusion, the number of molecules with more than four neighbors (that is, molecules with extra neighbors) is comparable to the number of molecules with exactly four neighbors. The number of intact hydrogen bonds is almost not affected by the increase of the number of molecules with higher coordination number. This indicates the presence of imperfections in the network what is supported by the presence of distortions in the H-O-H bond angles at higher densities (see Figs. 3a, and 5a). These new neighbor molecules provide extra oxygens that will share an hydrogen with another oxygen without forming an extra hydrogen bond. As a result, the interaction becomes weaker and might break. By a small rotation the molecule can connect to another molecule and by making small rotations, it is able to diffuse. This mechanism explains the coupling between rotation and translation. This fast diffusion is, however, limited to a certain range of densities that enable the molecule to make another bond once one had broken. At r_{min}, for instance, there is a very low probability to find these extra molecules in the neighborhood of a given molecule, and therefore the translational and rotational diffusion are slow.

REFERENCES

1. D. Eisenberg and W.J. Kauzmann. *The Structure and Properties of Water*. Clarendon, London, 1969.

2. P. Ball. *Life's Matrix: A Biography of Water*. Farrar Straus and Giroux, New York, 2000.

3. V. Brazhkin, S.V. Buldyrev, V.N Ryzhov, and H.E. Stanley, editors. *New Kinds of Phase Transitions: Transformations in disordered systems*, Dordrecht, 2002. NATO Advanced Research Workshop, Volga River, Kluwer.

4. O. Mishima and H.E. Stanley. Nature, 392, 164 (1998).

5. Martin Chaplin. Water structure and behavior, http://www.sbu.ac.uk/water.

6. E.W. Lang and H.-D. Lüdemann. *Angewandte Chemie, International Edition in English*, 21, 315 (1982).

7. R.C. Dougherty and L.N. Howard. Journal of Chemical Physics, 109, 7379 (1998).

8. J. R. Errington and P. G. Debenedetti. Nature, 409, 318 January (2001).

9. R.J. Speedy. Journal of Physical Chemistry, 86, 982 (1982).

10. R.J. Speedy. Journal of Physical Chemistry, 86, 3002 (1982).

11. R.J. Speedy. Journal of Physical Chemistry, 91, 3354 (1987).

12. P. H. Poole, F. Sciortino, U. Essmann, and H. E. Stanley. Nature, 360:324, november 1992.

13. P.H. Poole, F. Sciortino, U. Essmann, and H.E. Stanley. Physical Review E, 48, 3799 (1993).

14. F. Sciortino, P.H. Poole, U. Essmann, and H.E. Stanley. Physical Review E, 55, 727 (1997).

15. S. Harrington, R. Zhang, P.H. Poole, F. Sciortino, and H.E. Stanley. Physical Review Letters, 78, 2409 (1997).

16. H.E. Stanley and J. Teixeira. Journal of Chemical Physics, 73, 3404 (1980).

17. S. Sastry, P.G. Debenedetti, F. Sciortino, and H.E. Stanley. Physical Review E, 53, 6144 (1996).

18.   L.P.N. Rebelo, P.G. Debenedetti, and S. Sastry. Journal of Chemical Physics, 109, 626 (1998).

19.   S.B. Kiselev and J.F. Ely. Journal of Chemical Physics, 116, 5657 (2002).

20.   A. Geiger, F. H. Stillinger, and A. Rahman. Journal of Chemical Physics, 70, 4185 (1979).

21.   P.G. Debenedetti. Metastable Liquids. Princeton University Press, Princeton, 1996.

22.   J.T. Fourkas, D. Kivelson, U. Mohanty, and K.A. Nelson. *Supercooled Liquids: Advances and Novel Applications*. ACS Books, Washington, 1997.

23.   W. T. Pockman, J. S. Sperry, and J. W. O'Leary. Nature, 378, 715 December (1995).

24.   J. Jonas, T. DeFries, and D.J. Wilbur. Journal of Chemical Physics, 65, 582 (1976).

25.   F. X. Prielmeier, E. W. Lang, R. J. Speedy, and H. D. Lüdemann. Physical Review Letters, 1987.

26.   F.X. Prielmeier, E.W. Lang, R.J. Speedy, and H.-D. Ludemann. Berichte Bunsengeselschaft fur Physikalische Chemie, 92, 1111 (1988).

27.   M. R. Reddy and M. Berkowitz. Journal of Chemical Physics, 1987.

28.   F. Sciortino, A. Geiger, and H.E. Stanley. Nature, 354(november):218, november 1991.

29.   F. Sciortino, A. Geiger, and H. E. Stanley. Journal of Chemical Physics, 1992.

30.   L. A. Báez and P. Clancy. Journal of Chemical Physics, 101, 9837 (1994).

31.   S. Harrington, P.H. Poole, F. Sciortino, and H.E. Stanley. Journal of Chemical Physics, 107, 7443 (1997).

32.   P. Gallo, F. Sciortino, P. Tartaglia, and S.-H. Chen. Physical Review Letters, 76, 2730 (1996).

33.   F. Sciortino, P. Gallo, P. Tartaglia, and S.-H. Chen. Physical Review E, 1996.

34.   S.-H. Chen, P. Gallo, F. Sciortino, and P. Tartaglia. Physical Review E, 1997.

35.   F. Sciortino, L. Fabbian, S.-H. Chen, and P. Tartaglia. Physical Review E, 1997.

36.   T. Yamaguchi, S.-H. Chong, and F. Hirata. Journal of Chemical Physics, 119, 1021 (2003).

37. F. W. Starr, S. Harrington, F. Sciortino, and H. E. Stanley. Physical Review Letters, 1999.

38. F.W. Starr, F. Sciortino, and H.E. Stanley. Physical Review E, 60, 6757 (1999).

39. A. Scala, F.W. Starr, E.La Nave, F. Sciortino, and H.E. Stanley. Nature, 406, 166 (2000).

40. B. Guillot. Journal of Molecular Liquids, 101, 219 (2002).

41. A. G. Kalinichev. Reviews in Mineralogy and Geochemistry. Mineralogical Society of America, Washington DC, 2001.

42. H.J.C. Berendsen, J.R. Grigera, and T.P. Straatsma. Journal of Physical Chemistry, 91, 6269 (1987).

43. M.P. Allen and D.J. Tildesley. Computer Simulation of Liquids. Clarendon Press, Oxford, 1987.

44. J.P. Ryckaert, G. Ciccotti, and H.J.C. Berendsen. Journal of Computational Physics, 23, 327 (1977).

45. H.J.C. Berendsen, J.P.M. Postma, W.F.van Gunsteren, A. DiNola, and J.R. Haak. Journal of Chemical Physics,81, 3684 (1984).

46. 46 O. Steinhauser. Molecular Physics, 45, 335 (1982).

47. 47 Y. Yeh and C.-Y. Mou. Journal of Physical Chemistry B, 103, 3699 (1999).

48. 48 P.A. Netz, F.W. Starr, H.E. Stanley, and M.C. Barbosa. In New kinds of phase transitions: transformation in disordered substances, volume 81 of NATO Science Series II, page 417. NATO, 2002.

49. T. Head-Gordon and F. H. Stillinger. Journal of Chemical Physics, 98, 3313 (1993).

50. P.A. Netz, F.W. Starr, M.C. Barbosa, and H. Eugene Stanley. Physica A, 314, 470 (2002).

51. M. Mezei and D. L. Beveridge, Journal of Chemical Physics 74, 1981 (1981).

52. D.C. Rappaport. Molecular Physics, 50, 1151 (1983).

53. P. A. Netz, F. W. Starr, H. Eugene Stanley, and M. C. Barbosa. Journal of Chemical Physics, 115, 344 (2001).

54. M. Yamada, S. Mossa, H.E. Stanley and F. Sciortino Phys. Rev. Lett, 88, 195701 (2002).

55. H. E. Stanley, M.C. Barbosa, S. Mossa, P.A. Netz, F. Sciortino, F.W. Starr and M. Yamada Physica A, 315, 281 (2002).

56. H. E. Stanley, M.C. Barbosa, S. Mossa, P.A. Netz, F. Sciortino, F.W. Starr and M. Yamada *Water at positive and negative pressures* In *Liquids under Negative Pressures*, volume 84 of *NATO Science Series II*, NATO, 2002.

57. M. W. Mahoney and W. L. Jorgensen Journal of Chemical Physics, 112, 8910 (2000).

58. P. A. Netz, F. W. Starr, M. C. Barbosa, and H. Eugene Stanley. Journal of Molecular Liquids, 101, 159 (2002).

59. S. Ravichandran, A. Perera, M. Moreau, and B. Bagchi. Journal of Chemical Physics, 107, 8469 (1997).

60. H. Tanaka and I. Ohmine. Journal of Chemical Physics, 87 6128 (1987).

Chapter 4

COMPUTER SIMULATIONS OF CELL SORTING DUE TO DIFFERENTIAL ADHESION

Ying Zhang ,Gilberto[1] L. Thomas[2] ,Maciej Swat[3] ,Abbas Shirinifard[3] ,James A. Glazier[3]

[1]Cancer and Developmental Biology Laboratory, National Cancer Institute, Frederick, Maryland, United States of America

[2] Instituto de Fı́sica, Universidade Federal do Rio Grande do Sul, Porto Alegre, Brazil

[3] Biocomplexity Institute and Department of Physics, Indiana University, Bloomington, Indiana, United States of America

ABSTRACT

The actions of cell adhesion molecules, in particular, cadherins during embryonic development and morphogenesis more generally, regulate many aspects of cellular interactions, regulation and signaling. Often, a gradient of cadherin expression levels drives collective and relative cell motions generating macroscopic cell sorting. Computer simulations of cell sorting have focused on the interactions of cells with only a few discrete adhesion levels between cells, ignoring biologically observed continuous variations in expression levels and possible nonlinearities in molecular binding. In this paper, we present three models relating the surface density of cadherins to the net intercellular adhesion and interfacial tension for both discrete and continuous levels of cadherin expression. We then use then the Glazier-Graner-Hogeweg (GGH) model to investigate how variations in the distribution of the number of cadherins per cell and in the choice of binding model affect cell sorting. We find that an aggregate with a continuous variation in the level of a single type of cadherin molecule sorts more slowly than one with two levels. The rate of sorting increases strongly with the interfacial tension, which depends both on the maximum difference in number of cadherins per cell and on the binding model. Our approach helps connect signaling at the molecular level to tissue-level morphogenesis.

INTRODUCTION

The *cadherin* family of cell-adhesion membrane proteins plays a key role in both early and adult tissue morphogenesis [1]–[3]. Spatio-temporal variations in cadherin number and type help regulate many normal and pathological morphogenetic processes, including: neural-crest-cell migration [4], somite segmentation [5], [6], epithelial-to-mesenchymal transformations during tumor invasion and metastasis [7], [8], and wound healing [9], [10]. Many of these processes involve continuous variations in the expression level of a single type of adhesion molecule: During proximo-distal limb growth [11] and rostro-caudal body-axis elongation [12], adhesion gradients resulting from variations in the number of a single type of adhesion molecule may maintain cells› relative positions. *In vitro* and in experiments *in vivo*, when cells from different domains of a limb are mixed together, they can sort out according to their original positions[11], [13]. In *Drosophila*, an adhesion gradient drives the oocyte towards the posterior follicle cell, which expresses the highest level of DE-cadherin [14]. A cell-cell adhesion gradient along the dorso-ventral axis directs lateral cell migration during zebrafish gastrulation [15]. Thus, understanding the role of cadherins in creating and stabilizing tissue structures, especially the role of continuous variation in the level of a single cadherin, is crucial to understanding embryonic morphogenesis.

Steinberg's *Differential Adhesion Hypothesis* (*DAH*) originated the idea that cell sorting can result from variations in cell-cell adhesivity [16]–[19]. Cell sorting depends on the effective molecular binding strength between opposing cadherins, which in turn depends on their types and expression levels in each cell and potentially the cells› internal biochemistry and cytoskeletal structures [20]. Both differences in expression levels of a single type of cadherin[18], [19] and differences in the types of cadherins expressed [19], [21] can lead to sorting.

The relation between forces at the molecular level (pairs of cadherins), cell level (cell-cell adhesion), tissue level (surface tension) and cell sorting is more complicated than the simple physics suggested by the DAH. Experimental measurements of cadherin binding employing a variety of approaches have obtained widely differing estimates of the per-cadherin pair-binding force, cell-cell adhesion force and surface tension at the tissue level [19], [22]–[25]. In some experiments, the scaling between cadherin expression levels and surface tension, as given by equation (7), is quadratic (see equation (9)) [23]; in others, the scaling between cadherin expression levels and the cell-cell adhesion force is linear (see equation (10)) [19]. The cadherin organization within the cell membrane and the underlying cytoskeleton also change over a period of hours after two cells come into contact [3], [26]–[28]. Bindings between cadherin pairs differ for cadherins in different conformational states [3], *e.g.*, cadherin reorganization into adhesive patches on the cell membrane due to both passive diffusion and interaction with the actin cytoskeleton [26]–[28] can greatly increase the effective binding strength per cadherin pair between two cells. Cluster formation depends on the proper functioning of the actin cytoskeleton, so actin-disrupting drugs like cytochalasin-D and latrunculin greatly decrease cell-cell adhesivity [29].

Multiple transcriptional and post-translational signaling cascades can regulate cadherin expression levels, localization and per-cadherin binding strengths [3], [30]. In turn, cadherin binding can modify gene expression [3]. This complexity obscures the role of the cadherin-binding force in cell sorting [25]. As a result, different classes of experiments on specific types of cadherin have led to at least four simplified cadherin-binding models: the *linear-zipper model*(*LZM*) based on experiments on N-cadherin [31]–[34], the *cis-dimer model* (*CDM*) (equation (8)) based on experiments on E-cadherin [35], the *trans-homophilic-bond model* (*THBM*) (equation (9)) based on experiments on C-cadherin [36], and the *saturation model* (*SM*) (equation (10)), based on the observation that, for both the CDM and THBM models, when the cadherin binding between cells saturates, the number of bonds depends on the cell with the minimum cadherin concentration.

This paper therefore proposes a simple framework to explore how homotypic cadherin binding at the molecular level could produce intercellular adhesion and eventually determine cell sorting at the tissue level. We neglect complex spatial and temporal changes in cadherin behavior, assuming that cadherin distributions are uniform and constant on the cell membrane and that adhesion-strength per molecular bond is also time-independent (*i.e.*, we assume no conformational changes in molecular structure during a simulation). We then explore how the sorting configuration and rate depend on a few essential parameters in our models. Compared to the rate of sorting for an aggregate with two levels of a single cadherin, simulations with more intermediate levels sort more slowly but the sorting rate is similar for aggregates with the same number of cadherin levels for all binding models. The speed of sorting increases strongly with the interfacial tension, which depends both on the maximum difference in number of cadherins per cell and on the binding model.

METHODS

Reaction-Kinetic Models of E-Cadherin Binding

The nature of cadherin-cadherin binding determines the way the cell-cell adhesion energy, depends on cells' cadherin surface densities, and thus the correct binding model to use in simulations of cell sorting. Since more recent mutagenesis studies do not support the linear-zipper model [3], we use the cis-dimer (*CDM*), the trans-homophilic-bond (*THBM*), and the saturation (*SM*) models to relate the cells' cadherin surface densities to the cell-cell adhesion energy.

The cis-dimer model (CDM) [35] assumes that cis-dimers first form on the surfaces of individual cells and that two dimers on apposing cells then bind together to form homophilic tetramers. Dimerization of monomers (*A* and *A* or *B* and *B*) on individual cells' surfaces to form dimers *A2* and *B2* has the form:

$$A + A \rightleftharpoons A2; \; B + B \rightleftharpoons B2. \tag{1}$$

Similarly, when the trans-tetramer *A2B2* forms between dimers (*A2* and *B2*) on two apposing cells, the reaction has the form:

$$A2 + B2 \rightleftharpoons A2B2. \tag{2}$$

We assume that the cadherin concentrations on the cells' surfaces are constant and that we can apply the Law of Mass Action. Dimerization and tetramerization quickly equilibrate if K_D and K_T, the *equilibrium dimerization* and *equilibrium tetramerization dissociation constants* are large and the cadherin concentrations, $C_A = N_A/(S_A h)$ and $C_B = N_B/(S_B h)$, are lower than the dissociation constants [37]. Here N_A and N_B are the number of cadherin molecules distributed on the cell surfaces S_A and S_B, respectively, and h is the amplitude of cadherin fluctuations normal to the cells' surfaces. In this case, the total number of tetramers is less than the number of dimers, which in turn is less than the number of monomers. Then, the equilibrium concentration of tetramers in the CDM is, approximately,

$$[A2B2] = C_A^2 C_B^2/(K_D^2 K_T) = k_T N_A^2 N_B^2, \tag{3}$$

where $k_T = (K_D^2 K_T (S_A S_B h^2)^2)^{-1}$ is the *tetramer effective equilibrium constant*.

According to the trans-homophilic-bond model (THBM) [36], cadherins bind individually between cells, so the concentration of bound pairs is given by:

$$[A2B2] = C_A C_B/K_D = k_D N_A N_B, \tag{4}$$

where $k_D = (K_D S_A S_B h^2)^{-1}$ is the *dimer effective equilibrium constant*.

Finally, for the saturation model (SM), which applies for strong clustering of cadherins, or large differences in the number of molecules per cell, the concentration of bound cadherin pairs is given by

$$[A2B2] = \min\{C_A, C_B\} = k_M \min\{N_A, N_B\}, \tag{5}$$

where $k_M = ((S_A | S_B) h^2)^{-1}$ is the *effective equilibrium constant* and the surface $S = S_A | S_B$ corresponds to the smaller of C_A or C_B.

We relate the concentration of cadherin pairs to the cell-cell *intercellular adhesion energy density* due to cadherin binding via the relation:

$$J(N_A, N_B) = [A2B2]\Delta g + c, \tag{6}$$

where Δg is the cadherin-cadherin-binding free-energy per cadherin bond [37], which is negative, since bond formation releases energy, and where c is the *energy density due to adhesion unrelated to cadherins* [19].

The *interfacial-tension density* over the contact area between two cells expressing different numbers of a single type of cadherin is defined [38], [39] as:

$$\gamma_{A,B} = [J(N_A, N_A) + J(N_B, N_B)]/2 - J(N_A, N_B)$$

$$= (([A2] + [B2])/2 - [AB])\Delta g \tag{7}$$

For the three models just listed, equations (3–5), we have

$$:\gamma_{CDM} = -k_T(N_A^2 - N_B^2)^2 \Delta g/2, \tag{8}$$

$$\gamma_{THBM} = -k_D(N_A - N_B)^2 \Delta g/2, \tag{9}$$

$$\gamma_{SM} = -k_M(N_A - N_B)\Delta g/2, \text{ for } N_A > N_B. \tag{10}$$

Glazier-Graner-Hogeweg Simulations of Cell Sorting

To simulate cell sorting due to cell-cell adhesion, we used the *Glazier-Graner-Hogeweg* model (*GGH*) [40] (also known as the *Cellular Potts Model* [38], [39]). The GGH is a multi-cell, lattice-based model, which uses an *effective energy*, H, to describe the behavior of cells, for instance, due to cell-cell adhesion. GGH simulations agree quantitatively with simple cell-sorting and other experiments [41]–[49].

Cells in the GGH are extended domains of pixels (on a regular lattice, denoted $\vec{i}$), which share the same *cell index*, $\sigma(\vec{i})$. The effective energy governs how the lattice evolves as cells attempt to displace other cells by extending their pseudopodia [50]. At each step, we select a lattice site $\vec{i}'$ and change its index into the index of a neighboring lattice site $\vec{i}$ with probability:

$$P(\sigma(\vec{i}') \to \sigma(\vec{i})) = \begin{cases} \exp(-\Delta H/T) & \text{if } \Delta H > 0; \\ 1 & \text{if } \Delta H \leq 0, \end{cases} \tag{11}$$

where ΔH is the energy gain from the change and T is the *intrinsic cell motility* corresponding to membrane fluctuations resulting from cytoskeleton fluctuations. If the lattice has ϱ pixels, we define one *Monte Carlo Step* (*MCS*) to be ϱ displacement attempts.

For a two-dimensional simulation of an aggregate containing cells expressing varying levels of a single type of cadherin, we assume that: (1) The effective energy between cells is due to cell-cell adhesion. (2) The cells have fixed and identical target volumes, membrane areas, and intrinsic motilities. (3) Cells do not grow, divide or die. (4) Cells are isotropic, so cadherins are uniformly distributed on the cell membrane and the cadherin concentration is constant in time. With these assumptions, the effective energy

$$H = \sum_{\substack{\vec{i},\vec{i}' \\ \text{neighbors}}} \left\{ J_0 + J\left(N_{\sigma(\vec{i})}, N'_{\sigma'(\vec{i}')}\right)\left(1 - \delta_{\sigma(\vec{i}),\sigma'(\vec{i}')}\right) \right\}$$

$$\text{is:} \qquad + \sum_{\sigma} \lambda (V(\sigma) - V_t)^2, \qquad (12)$$

where, J_0 is the energy per unit contact area between two cells in the absence of cadherin binding, which may be positive since such cells may not cohere. $J(N_{\sigma(\vec{i})}, N_{\sigma(\vec{i}')})$ is the adhesion-energy per unit contact area between cells σ and σ' expressing N and N' adhesion molecules, respectively. This term is always negative, since forming cadherin bonds decreases the effective energy. Sums go up to fourth nearest neighbors on a square lattice. λ, $V(\sigma)$, and V_t are the volume elasticity, actual volume and target volume of cell σ, respectively. $\delta_{\sigma(\vec{i}),\sigma(\vec{i}')}$ is the usual Kronecker delta function.

Each cell expresses a specific number of cadherins. The cell-cell adhesion energy relates to N and N' according to equation (6) together with equations (3), (4) or (5). Since we can rescale the energy by the intrinsic cell motility, we are free to pick the energy scale and set $\Delta g = -1$.

The relative strengths of cell-cell adhesions result in net forces which act on each cell. Depending on the relative hierarchy of cell-cell adhesive interactions the generated forces can either drive or suppress cell sorting. Equation (13) is the condition for the sorting to occur.

Why does sorting occur for most of the conditions that we consider in this paper? For two cadherin levels with $N_A > N_B$, complete sorting requires that the less cohesive cell type wet the more cohesive cell type [39]

$$J(N_A,N_A) < [J(N_A,N_A) + J(N_B,N_B)] /$$
$$:2 < J(N_A,N_B) < J(N_B,N_B). \qquad (13)$$

Since $-\frac{N_A^2 + N_B^2}{2} < -N_A N_B$ for the THBM, $-\frac{N_A^4 + N_B^4}{2} < -N_A^2 N_B^2$ for the CDM, and $-\frac{N_A + N_B}{2} < -\min\{N_A, N_B\}$ for the SM, the binding energies all satisfy the sorting condition. Therefore, cells should sort for all three binding models. Even cells with a continuous distribution of cadherin levels satisfy the sorting inequality, so cells with fewer adhesion molecules envelop cells with more adhesion molecules, which sort towards the center of the aggregate, creating an adhesion gradient, decreasing from the center to the periphery (Figure 1E), with a small amount of local mixing due to intrinsic cell motility. As mentioned above, sorting is a simple mechanism for cells to reach and maintain their positions during morphogenesis, *e.g.*, during limb outgrowth, in which cells maintain both their antero-posterior and proximo-distal positions through differential adhesion.

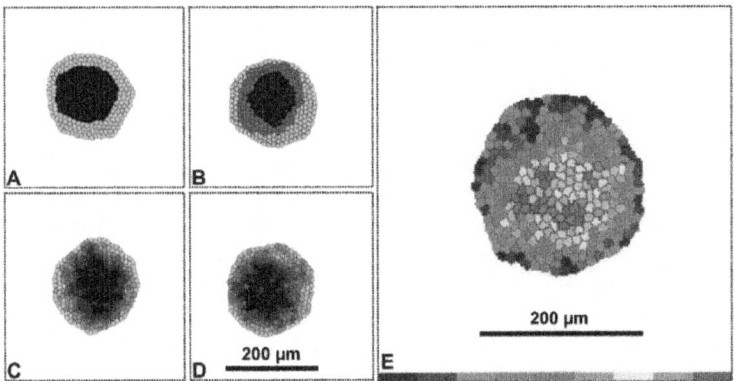

Figure 1: Typical simulated sorted configurations for aggregates of cells for the trans-homophilic-bond model (THBM).

All images shown at time t=999,000 MCS. In A–D, the gray-scale represents the cadherin-expression level. The darkest color (gray level=0) represents the highest cadherin-expression level. The lightest color (gray level=200) represents the lowest cadherin-expression level. The cell culture medium is white (gray level=255). In (E), HSV colors represent the expression levels, ($H = [1-(N-N_{min})/(N_{max}-N_{min})]255$, $S = 255$, $V = 255$), where N is the cadherin-expression level, and N_{min} and N_{max} are the minimum and maximum cadherin-expression levels, respectively. Red ($H = 0$) is the highest expression level, blue ($H = 255$) the lowest expression level. The cell culture medium is white. Sorting for: (A) 2 levels. (B) 3 levels. (C) 5 levels. (D) 9 levels. (E) Continuous levels. Cadherin expression ranges from $N_{min} = 1$ to $N_{max} = 23$. In all simulations, $T = 20$ and $\lambda = 25$.

In an ideal, fully-sorted configuration, cells expressing the higher levels of cadherins will cluster together and round up into a solid sphere, surrounded by successive spherical shells of cells expressing successively lower levels of cadherins. To monitor the progress of cell sorting in our simulations, we define the heterotypic boundary length (*HBL*), the total contact length between cells with different cadherin levels, measured in pixels:

$$L_h = \sum_{\overline{i,i'}\,\text{neighbors}} \left(1 - \delta(N_{\sigma(\overline{i})}, N'_{\sigma'(\overline{i'})})\right).$$

$$(14)$$

The simulations time evolution gradually minimizes L_h.

If cells express multiple cadherin levels, L_W, the heterotypic boundary length weighted by the energy differences between neighboring cells is a better metric for cell sorting. This weighted heterotypic boundary length (*WHBL*) is simply the total interfacial tension (equations (8–10)) multiplied by the lengths:

$$L_{W_{CDM}} = -k_T \sum_{\overset{i,i'}{\text{neighbors}}} \left(N^2_{\sigma(i)} - N'^2_{\sigma'(i')} \right)^2$$

$$\left(1 - \delta_{N_{\sigma(i)}, N'_{\sigma'(i')}} \right) \Delta g / 2, \tag{15}$$

$$L_{W_{THBM}} = -k_D \sum_{\overset{i,i'}{\text{neighbors}}} \left(N_{\sigma(i)} - N'_{\sigma'(i')} \right)^2$$

$$\left(1 - \delta_{N_{\sigma(i)}, N'_{\sigma'(i')}} \right) \Delta g / 2, \text{ and} \tag{16}$$

$$L_{W_{SM}} = -k_M \sum_{\overset{i,i'}{\text{neighbors}}} \left(N_{\sigma(i)} - N'_{\sigma'(i')} \right)$$

$$\left(1 - \delta_{N_{\sigma(i)}, N'_{\sigma'(i')}} \right) \Delta g / 2. \tag{17}$$

Different aggregates may have different maximum (initial) and minimum heterotypic boundary lengths (HBL) or weighted heterotypic boundary lengths (WHBL). To compare sorting in different aggregates, we normalize these lengths using the transformation:

$$l_{norm} = \frac{L(t) - L_{Min}}{L_{Max} - L_{Min}}, \tag{18}$$

where $L_{Max} = \max\{L(t)\}$, $L_{Min} = L_k$ or L_{theo}, and $L(t)$ is the HBL or WHBL at time t. $L_k = \min\{L(t)\}$ is the minimum value of HBL or WHBL over the typical simulation duration of 10^6 MCS. L_{theo} is the theoretical minimum HBL or WHBL for the fully sorted and rounded aggregate, assuming that the cells form perfect concentric rings with perimeters equal to $2\pi R$ (R is the real radius of the ring of cells from the center of the aggregate). Experimentally, this value is easily calculated with digital imaging analysis, which gives us the total area of each type of cell. The *sorting relaxation time*, τ, is the time at which the aggregate reaches its typical, maximally-sorted configuration. τ is defined via the relation:

$$l_{norm}(\tau) = \frac{l_{norm}(0)}{e},$$

(19)

The *sorting rate*, R_S, is the inverse of the sorting relaxation time:

$$R_S = \tau^{-1}.$$

(20)

Simulation Implementation

We first investigated sorting completeness for the trans-homophilic-bond model (THBM, equation (4)), with $k_D = 0.02$, as we moved from two levels of cadherin expression towards a continuous distribution of levels (two, three, five, nine and continuous levels) with the same range of cadherin numbers, [$N_{min} = 1$, $N_{max} = 23$]. The same range of cadherin expression numbers provides the same range of adhesion energies, independent of the number of levels.

We implemented our simulations using the open-source software package CompuCell3D (downloadable from http://www.compucell3d.org/) which allows rapid translation of biological models into simulations using a combination of CC3DML and Python scripting. We presented our simulation codes in Codes S1.

All our simulations for cell sorting use aggregates of 305 cells, close to the size of a 2D section of the 3D aggregates experimentally studied by Armstrong, Steinberg and others [18], [41],[51], which are about 200 microns in diameter. Each cell has a 25-pixel target volume, which sets the lattice length scale to approximately 2 microns per pixel. We begin with a circular-disk aggregate with cells randomly assigned cadherin expression numbers, with each allowed number having equal probability. Each simulation uses $T = 20$ and runs for 10^6 MCS, to allow for complete sorting for continuous variations of cadherin expression over the range [1, 23]. We set $\lambda = 25$, which allows patterns to evolve reasonably fast without large cell-volume or cell-surface-area fluctuations. Changing λ around this value does not greatly affect the relaxation of cells' shapes and positions. We further set $J_0 = 16$ (in equation (12)) for all simulations. For different cadherin binding models and for the cadherin expression range [1, 23], we choose the values of k_T, k_D and k_M (according to equation (3–5)), so that cells neither pin to the lattice nor dissociate.

RESULTS

Figures 1A–E show final aggregates for cells expressing discrete or continuous levels of cadherins. Cells with higher expression (darker gray in Figures 1A–D, red in Figure 1E) assume more central positions, while cells with lower expression (lighter gray in Figures 1A–D, blue in Figure 1E) move to the periphery. For multiple discrete levels, cells follow a sorting hierarchy [17]; each layer of cells has a given expression number and surrounds the layer of cells with the next-higher level. For continuous levels, expression numbers decrease continuously from the center to the periphery of the aggregate (Figure 1E).

We investigated the evolution of the effective energy and the heterotypic boundary length (HBL)/weighted heterotypic boundary length (WHBL) for the THBM (equation (16)) in three cases:

1. Cells with different numbers of levels of cadherin expression, but the same range between maximum and minimum expression number.

2. Cells with different ranges between maximum and minimum expression number but with the same number of levels (two, for simplicity).

3. Cells with different motilities, but with the same cadherin levels.

We also investigated:

1. Cells with different cadherin binding models, but the same range between maximum and minimum expression number for two, five, nine and continuous levels.

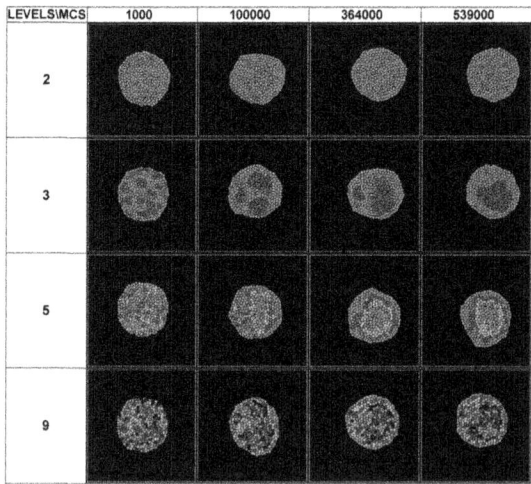

Figure 2: Simulation snapshots for aggregates with differing numbers of cadherin levels, with the same maximum to minimum expression range [1, 23], for the THBM.

Figure 2 shows sets of snapshots of simulations for cell aggregates with the THBM (equation (4) with $k_D = 0.02$, $T = 20$, and $\lambda = 25$) with cells expressing two [1, 23], three [1, 12, 23], five [1, 6.5, 12, 17.5, 23], nine [1, 3.75, 6.5, 10.25, 12, 14.75, 17.5, 20.25, 23] cadherin levels. The corresponding animations are in: Movie S1, Movie S2, Movie S3, and Movie S4.

Figure 3A shows the evolution of the effective energy H for the cell aggregates presented inFigure 2, and for cell aggregates with continuous cadherin levels in the range [1, 23] calculated using the THBM (equation (4) with $k_D = 0.02$, $T = 20$, and $\lambda = 25$). Figures 3B and 3Cillustrate the evolution of the normalized weighted heterotypic boundary length (NWHBL) for the cell aggregates in Figure 3A, setting $L_{min} = L_k$ and $L_{min} = L_{theo}$, respectively. Aggregates with two or three levels sort quickly, while those with more levels take more time to sort (Figure 2D).

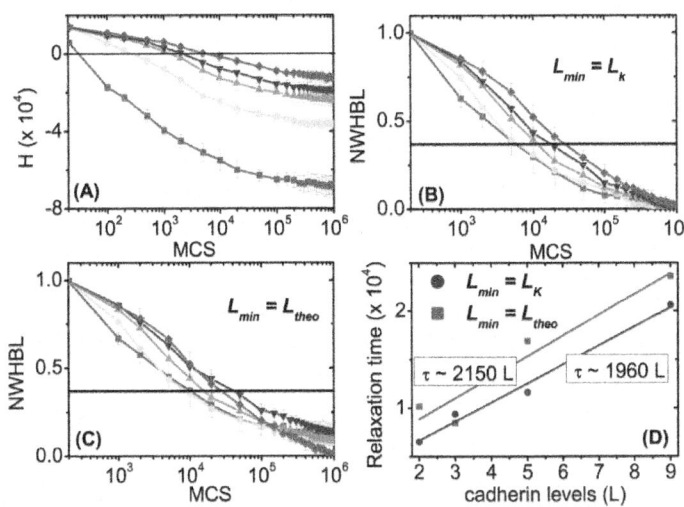

Figure 3: Evolution of the effective energies (H) and normalized weighted heterotypic boundary lengths (NWHBL) for aggregates with differing numbers of cadherin levels, with the same maximum to minimum expression range [1, 23], for the THBM.

(A)–(C)■—2 levels; •– 3 levels; ▲ – 5 levels; ▼ – 9 levels; ◆ – continuous levels. The black horizontal lines mark 1/e. (A) Evolution of H. (B)–(C) Evolution of the NWHBL for the simulations in (A), with $L_{min} = L_k$ in (B) and $L_{min} = L_{theo}$ in (C). (D) Relaxation time *vs.* number of levels. ▪: $L_{min} = L_{theo}$. •: $L_{min} = L_k$. The graphs are calculated from ten simulation replicas.

Figure 4A shows the evolution of the effective energy H for aggregates with two cadherin levels, but different expression ranges: [1, 12], [1, 14.75], [1, 17.50], [1, 20.25], [1, 23], [12, 23], and [19.62, 23], also calculated using the THBM (equation (4) with $k_D = 0.02$, $T = 20$, and $\lambda = 25$). Figures 4B and 4C show the evolution of the NWHBL for the same aggregates, using $L_{min} = L_k$ and $L_{min} = L_{theo}$, respectively. Sorting is quickest ($\tau \simeq 14,000$ MCS) for aggregates with the widest cadherin expression range [1, 23], and is slowest (no complete sorting, $\tau = \infty$) for aggregates with the smallest expression range [19.62, 23].

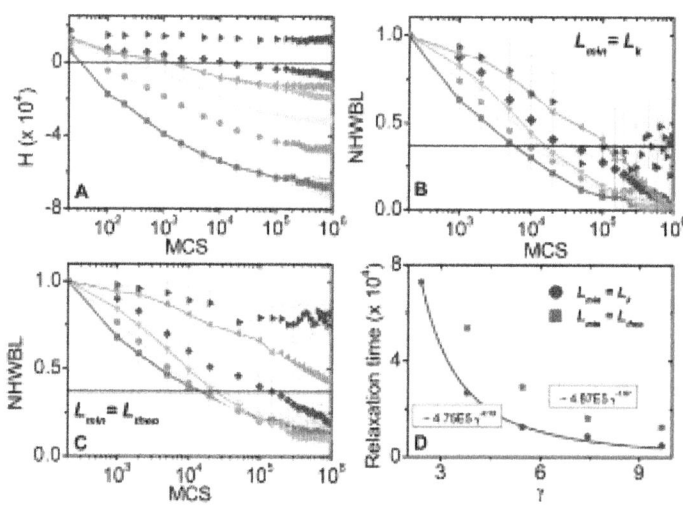

Figure 4: Evolution of the effective energies (H) and normalized weighted heterotypic boundary lengths (NWHBL) for aggregates expressing 2 cadherins levels modeled with the THBM.

Expression ranges: ■ – [1, 23]; • – [1, 20.25]; ▲ – [1, 17.50]; ▼ – [1,14.75]; ◄ – [12, 23]; ◆ – [1, 12]; ► – [19.62, 23]. Black solid horizontal lines mark 1/e. (A) Evolution of H. (B)–(C) Evolution of NWHBL for the simulations in (A) with: (B) $L_{min} = L_k$, and (C) $L_{min} = L_{theo}$. (D) Relaxation time *vs.* interfacial tension γ. Dots – simulation, and Lines– fitting curves ax^b. • – $L_{min} = L_k$; ■ – $L_{min} = L_{theo}$. The error bars in the graphs are calculated from ten simulation replicas.

doi:10.1371/journal.pone.0024999.g004

According to the theory of phase separation in liquids, the sorting rate for simple fluids is proportional to the interfacial tension divided by the viscosity' [52]. A similar relationship may hold for cell sorting [53]. Figure 4D plots the sorting relaxation time against the interfacial tension (equation 9) for the simulated aggregates in Figure 4A, and a power law (of form $\tau = a\gamma_{THBM}^b$, with a and b constants), fitting for both the cases $L_{min} = L_k$ and $L_{min} = L_{theo}$, respectively:

$$\tau_k = 4.75 \times 10^5 \gamma_{THBM}^{-2.13} \quad \text{and} \tag{21}$$

$$\tau_{theo} = 4.87 \times 10^5 \gamma_{THBM}^{-1.67}. \tag{22}$$

The fitting is reasonable, since for $L_{min} = L_{theo}$, the adjusted coefficient of determination $R^2 = 0.89$, and for $L_{min} = L_k$, $R^2 = 0.98$, suggesting that the sorting relaxation time and interfacial tension may obey an approximate power law with an exponent $b \simeq -2$.

In Figure 5 we compare the evolution of the effective energy H and of the NWHBL for the different cadherin binding models (CDM, THBM, and SM), with two, five, nine and continuous cadherin levels (the same levels as in Figure 3). We chose the effective equilibrium constants (see equations (8)–(10)), $k_D = 0.02$, $k_T = 0.000038$, and $k_M = 0.46$, so the cell-cell adhesion energies fell in the same range, excluding changes in cell sorting rates due to differences in these ranges. Figure 5D3 shows that as the number of expression levels increases from 2 to 5 to 9, the relaxation time increases for each model.

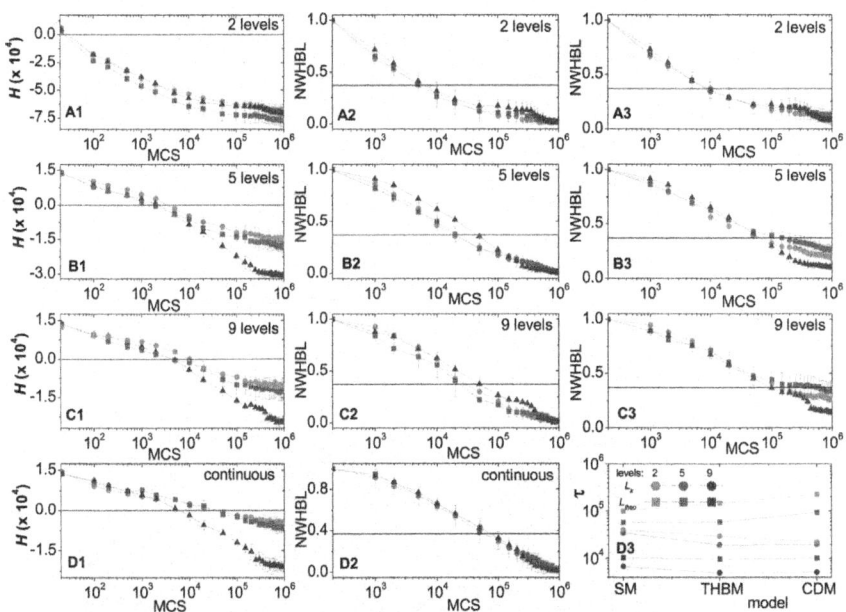

Figure 5: Evolution of the effective energies (H) and normalized weighted heterotypic boundary lengths (NWHBL) for aggregates with 2, 5, 9 or continuous cadherin levels using CDM, THBM or SM for the same expression range [1, 23].

In (A1)–(A3), (B1)–(B3), (C1)–(C3), and (D1)–(D2), Red lines · – CDM; Green lines • – THBM; Blue lines ▲ – SM. In (A2), (A3), (B2), (B3) and (C2)–(C4), the time at which the heterotypic boundary length of a given simulation crosses the horizontal black line is defined as its relaxation time. In (A2), (B2), (C2) and (D2) $L_{min} = L_k$. In (A3), (B3) and (C3) $L_{min} = L_{theo}$. (A1), (B1), (C1) and (D1) Evolution of the H for aggregates with cells expressing 2, 5, 9 and continuous cadherin levels respectively. (A2), (B2), (C2) and (D2) Evolution of NWHBL for the aggregates in (A1), (B1), (C1) and (D1), respectively, with $L_{min} = L_k$. (A3), (B3) and (C3) Evolution of the NWHBL for the aggregates in (A1), (B1) and (C1) respectively, with $L_{min} = L_{theo}$. (D3) Relaxation time *vs.* bond model for different cadherin expression levels. **Blue –** 2 levels; Red – 5 levels; Green – 9 levels. Circles – $L_{min} = L_{theo}$. Squares – $L_{min} = L_k$. The error bars in the graphs are calculated from ten simulation replicas.

doi:10.1371/journal.pone.0024999.g005

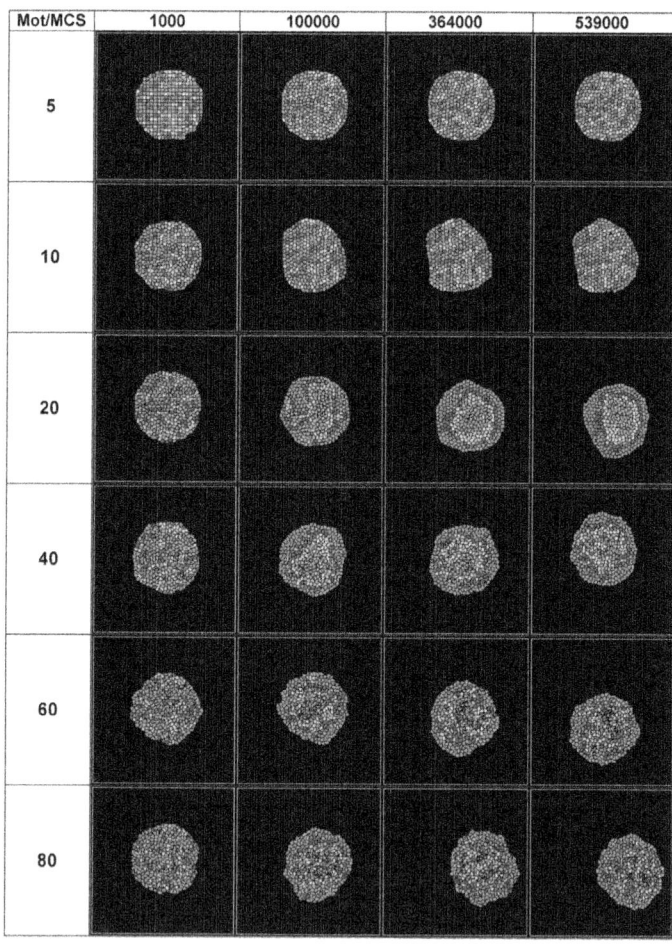

Figure 6: Simulation snapshots for aggregates with five levels [1, 6.5, 12, 17.5, 23] of cadherins and different cell motilities (5, 10, 20, 40, 60, 80), for the THBM.

For different models with the same cadherin expression levels, for two-level aggregates (Figure 5A1), sorting times are equal, as we expect because equations (8–10) give almost identical interfacial tensions. For aggregates with five and nine cadherin levels (Figures 5B1 and 5C1), sorting is more rapid for the saturation model (SM) and slowest for the trans-homophilic-bond model (THBM). The average minimum WHBLs are largest for the SM, but are the same for the cis-dimer model (CDM) and THBM. Since the weighted heterotypic boundary length (WHBL) is actually the interfacial tension, it is the main factor which determines the sorting rate.

Figure 6 shows sets of snapshots of simulations for cell aggregates with the THBM (equation (4) with $k_D = 0.02$, $T = 20$, and $\lambda = 25$) with five cadherin levels [1, 6.5, 12, 17.5, 23] and different cell motilities: 5, 10, 20, 40, 60, and 80. The corresponding animations are in: Movie S5, Movie S6, Movie S3, Movie S7, Movie S8, and Movie S9.

Figure 7 shows the effect of cell motility on the evolution of the effective energy and normalized WHBL for aggregates with two cadherin levels using the THBM (with $k_D = 0.02$ and $\lambda = 25$).Figures 7A and 7B show the evolution of the effective energy for fixed λ. If the cell motility is very low ($T = 5$), cells pin before reaching their lowest-energy positions and sorting is slow. As the motility grows, the aggregates sort faster (Figure 5A). However, if the cell motility is too large ($T = 60$ and $T = 80$), sorting is rapid but remains incomplete (Figures 7B, 7C and 7D).

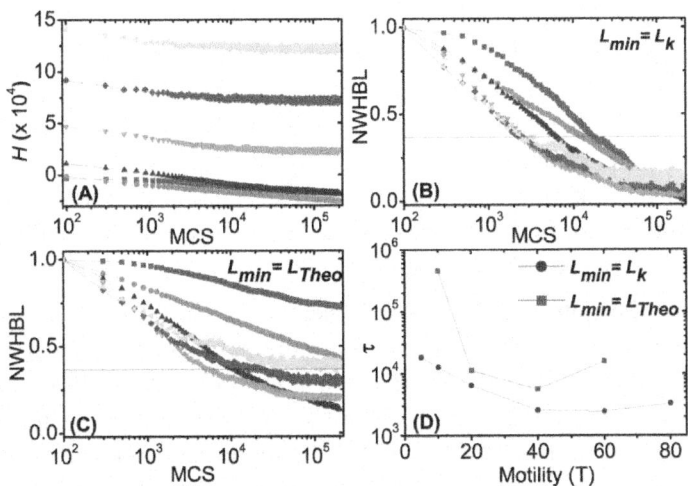

Figure 7: Evolution of the effective energies (H) and normalized weighted heterotypic boundary lengths (NWHBL) for aggregates with 5 cadherin levels and the same maximum to minimum expression range [1, 23] using the THBM with different motilities.

▪ − 5; • − 10; ▲ − 20; ▼ − 40; ◆ − 60; ◄ − 80. (A) Evolution of H. In (B) and (C) the time at which the heterotypic boundary length of a given simulation crosses the horizontal black line is defined as its relaxation time τ. In (B) $L_{min} = L_k$ and in (C) $L_{min} = L_{theo}$. (D) Relaxation time vs. relative cell motilities. ▪ − $L_{min} = L_{theo}$; • − $L_{min} = L_k$. The error bars in the graphs are calculated from ten simulation replicas.

When cells' expression of cadherin varies continuously, sorting still occurs, but more slowly than for discrete expression levels. The final configuration is imperfectly sorted since the intrinsic cell motility can overcome small differences in adhesion energy due to local missorting. The sorting rate depends on the interfacial tension rather than directly on the expression levels or the cadherin-binding model. Again, insufficient or excessive motility prevents complete sorting.

From the considerations above we can say that, although individually the sorting kinetics in aggregates with each binding model are sensitive to the number of cadherin levels and the energy range, all models have similar global behaviors. For each model, sorting is always faster for smaller numbers of cadherin levels, independent of the energy expression range. The dependence of sorting time and completeness on the number of cadherin levels is also similar for the three models, although the SM model seems to sort slightly faster and more completely for large numbers of cadherin levels. In the absence of experiments determining the model to use, the SM is computationally more efficient for larger aggregates.

Our results could be checked by experiments controlling cadherin expression. *E.g.* we could transfect a GFP-cadherin plasmid construct into normally non-adherent CHO cells, so the amount of cadherin in each cell would be proportional to its fluorescence intensity. For discrete levels we could use multiple fluorescent tags. Cotransformation with a nuclear-targeted fluorescent protein of a different color would allow real-time cell tracking to determine cell motilities and positions.

Using the interfacial boundary length as a measure of sorting is experimentally inconvenient because current automated image segmentation cannot accurately extract the interfacial lengths from a stack of images. Instead, measuring the autocorrelation of the intensity in experimental and simulation image stacks would be much simpler. To represent a nuclear-targeted label in our simulations we could place a dot at each cell›s center of mass with an intensity proportional to the cell›s the number of cadherins. To represent cytoplasmic labeling, we could fill the entire cell volume with an intensity corresponding to the cadherin level and similarly for membrane labeling, we could label the cell›s contour.

An alternative measure of sorting would use a clustering algorithm to track the number and size of homotypic cell clusters. This approach is straightforward in CC3D and relatively easy to implement in experiments using K-Means or K-Median clustering algorithms, as described in[54]. Figure 8 shows an example of this procedure. We have used a bigger aggregate, with about 5000 cells in order to have a reasonable statistics.

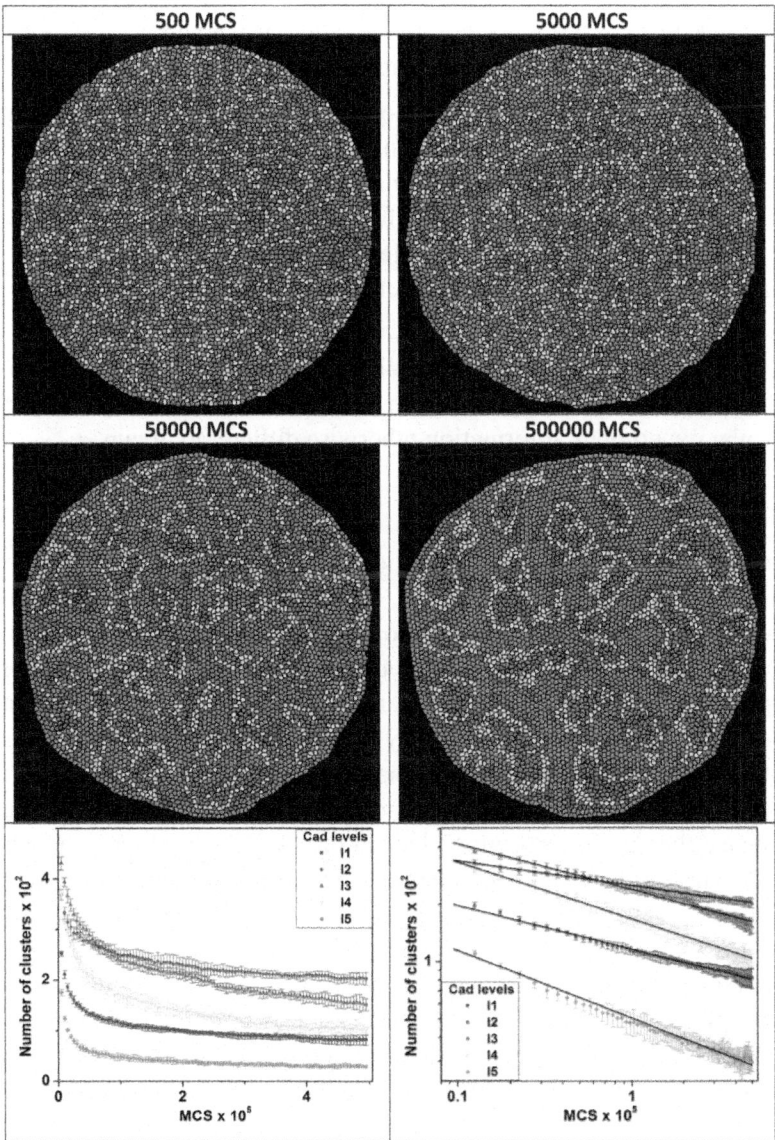

Figure 8: Clustering dynamics.

The cells have five levels of cadherins (as in Figure 6) and initially they are randomly distributed within the aggregate (top left snapshot). In our simple clustering algorithm, cells that express the same amount of cadherin and are in direct contact belong to the same cluster. The initial small clusters rapidly coalesce and form large clusters (top right and second row snapshots). The graphs at the bottom row show that the clustering rates decrease with time (left graph) and that they are adequately fitted by a power law of ab^t, as can be seen from black lines in the *log-log* graph at right. Mean and error bars for these graphs are calculated from six simulation replicas.

First and second rows: snapshots taken from a 5000 cell aggregate simulation with five levels of cadherins [1(l1), 6.5(l2), 12(l3), 17.5(l4), 23(l5)] showing the dynamics of cluster formation. Bottom row: the left graph shows the evolution of the number of cluster for each cadherin level. The *log-log* graph (right) shows that the dynamics is adequately fitted by a power law of at^b, as indicated by the black lines. The error bars in the graphs are calculated from six simulation replicas.

Comparing any of these bulk cell-sorting measures for experiments and simulations would allow us to infer the specific binding mechanism in a particular experiment, information otherwise difficult to obtain.

DISCUSSION

At the beginning of a particular developmental phase, patterns of gene expression are often fuzzy initially, then gradually become distinct. Both changing cell identity and cell movement are possible mechanisms for refining initially fuzzy expression patterns or for fixing transient patterns of morphogens. Glazier *et al.* 2008 [49] and Watanabe *et al.* 2009 [55], found that, during somite segmentation, the fuzzy boundary formed by cells, disregarding positional cues and differentiating inappropriately, can reorganize to form a sharp boundary due to cell motility and differential adhesion. The sorting rate, and hence the rate of patterning, depend on the interfacial tensions, which in turn depend on the range of cadherin expression, equilibrium constants and free energies of cadherin bonds (see equations (8–10)). These mechanisms may act in parallel with, or coordinate with, other morphogenic mechanisms, such as Turing-type reaction-diffusion instabilities or Wolpertian threshold-based positional coding. Adhesion mechanisms act as an effective low-pass filter, reducing the effect of stochasticity in gene expression. During development, signaling cascades modulate cadherin expression. Because cell sorting is slow compared to fluctuations in gene-expression levels and because sorting rectifies noise into a stable gradient, transient fluctuations in cadherin expression will not change final morphology, increasing developmental robustness.

To provide better links/interplay between computer simulations and biological experiments, we would suggest carrying out measurements of the following key parameters [19], [22]–[25]: individual cell motilities, positions, contours and boundary lengths and tissue and single-cell level adhesion protein expression, elasticity and viscosity. While not always accessible, measurements of one or more adhesion-related parameters such as interfacial tension between cell aggregates, cell-cell adhesion forces or energies, molecular binding forces or energies and molecular binding and junction-formation kinetics would facilitate constructions of more realistic computer simulations. In particular, the ability to measure and then model temporal variation of adhesion related parameters is essential for simulations of complex developmental phenomena such as somitogenesis, limb growth, *etc.*... Therefore future measurements should concentrate on dynamics of intra and inter-cellular mechanisms (*e.g.*intercellular signaling and regulatory networks) related to cellular adhesion [3], [26]–[28].

Our studies based on the Glazier-Graner-Hogeweg model, investigated how homotypic cadherin binding at the molecular level affects cell-cell adhesion and determines cell sorting speeds at the tissue level. We have used three different microscopic models of cadherin-binding for discrete and continuous levels. The three binding mechanisms lead to similar cell-sorting behavior, although the saturation binding model is somewhat faster for larger aggregates with more cadherin levels. Sorting speed decreases with increasing numbers of cadherin levels. For classical sorting with two cadherin levels, sorting speed increases with the ratio between the two levels. Additionally, in each case a single optimum value for the cell motility results in the fastest sorting. Cell motilities above or below the optimum sort more slowly.

AUTHOR CONTRIBUTIONS

Conceived and designed the experiments: YZ JAG GLT AS. Performed the experiments: YZ GLT AS. Analyzed the data: GLT YZ JAG. Contributed reagents/materials/analysis tools: YZ GLT MS AS. Wrote the paper: GLT YZ JAG MS. Helped with coding: MS.

REFERENCES

1.   Radice GL, Rayburn H, Matsunami H, Knudsen KA, Takeichi K, et al. (1997) Developmental defects in mouse embryos lacking N-cadherin. Dev Bio 181: 64–78.

2.   Price SR, de Marco Garcia NV, Ranscht B, Jessell TM (2002) Regulation of motor neuron pool sorting by differential expression of type II cadherins. Cell 109: 205–216.

3. Gumbiner BM (2005) Regulation of cadherin-mediated adhesion in morphogenesis. Nature Rev Mol Cell Bio 6: 622–634.

4. Xu X, Li WE, Huang GY, Meyer R, Chen T, et al. (2001) Modulation of mouse neural crest cell motility by N-cadherin and connexin 43 gap junctions. J Cell Bio 154: 217–230.

5. Linask KK, Ludwig C, Han MD, Liu X, Radice GL, et al. (1998) Ncadherin/catenin-mediated morphoregulation of somite formation. Dev Bio 202: 85–102.

6. Horikawa K, Radice GL, Takeichi M, Chisaka O (1999) Adhesive subdivisions intrinsic to the epithelial somites. Dev Bio 215: 182–189.

7. Takeichi M (1993) Cadherins in cancer: implications for invasion and metastasis. Curr Op Cell Bio 5: 806–811.

8. Berx G, Van Roy F (2001) The E-cadherin/catenin complex: an important gatekeeper in breast cancer tumorigenesis and malignant progression. Breast Cancer Res 3: 289–293.

9. Bement WM, Forscher P, Mooseker MS (1993) A novel cytoskeletal structure involved in purse string wound closure and cell polarity maintenance. J Cell Biol 121: 565–578.

10. Lorger M, Moelling K (2006) Regulation of epithelial wound closure and intercellular adhesion by interaction of AF6 with actin cytoskeleton. J Cell Sci 119: 3385–3398.

11. Yajima H, Yoneitamura S, Watanabe N, Tamura K, Ide H (1999) Role of N-cadherin in the sorting-out of mesenchymal cells and in the positional identity along the proximodistal axis of the chick limb bud. Dev Dyn 216: 274–284.

12. Bitzur S, Kam Z, Geiger B (1994) Structure and distribution of N-cadherin in developing zebrafish embryos: morphogenetic effects of ectopic over-expression. Dev Dyn 201: 121–136.

13. .Omi M, Anderson R, Muneoka K (2002) Differential cell affinity and sorting of anterior and posterior cells during outgrowth of recombinant avian limb buds. Dev Biol 250: 292–304.

14. Godt D, Tepass U (1998) Drosophila oocyte localization is mediated by differential cadherin-based adhesion. Nature 395: 387–391.

15. Von der Hardt S, Bakkers J, Inbal A, Carvalho L, Solnica-Krezel L, et al. (2007) The Bmp gradient of the zebrafish gastrula guides migrating lateral cells by regulating cell-cell adhesion. Curr Bio 17: 475–487.

16. .Steinberg MS (1963) Reconstruction of tissues by dissociated cells. Some morphogenetic tissue movements and the sorting out of embryonic cells may have a commonexplanation. Science 141: 401–408.

17. .Steinberg MS, Wiseman LL (1972) Do morphogenetic tissue rearrangements require active cell movements? The reversible inhibition of cell sorting and tissue spreading by cytochalasin B. J Cell Bio 55: 606–615.

18. .Steinberg MS, Takeichi M (1994) Experimental specification of cell sorting, tissue spreading, and specific spatial patterning by quantitative differences in cadherin expression. Proc Nat Ac Sc USA 91: 206–209.

19. Foty RA, Steinberg MS (2005) The differential adhesion hypothesis: a direct evaluation. Dev Bio 278: 255–263.

20. Friedlander DR, Mege RM, Cunningham BA, Edelman GM (1989) Cell sorting-out is modulated by both the specificity and amount of different cell adhesion molecules (CAMs) expressed on cell surfaces. Proc Nat Ac Sc USA 86: 7043–7047.

21. .Niessen CM, Gumbiner BM (2002) Cadherin-mediated cell sorting not determined by binding or adhesion specificity. J Cell Biol 156: 389–399.

22. Baumgartner W, Hinterdorfer P, Ness W, Raab A, Vestweber H, et al. (2000) Cadherin interaction probed by atomic force microscopy. Proc Nat Ac Sci USA 97: 4005–4010.

23. Chu YS, Thomas WA, Eder O, Pincet E, Perez E, et al. (2004) Force measurements in E-cadherin-mediated cell doublets reveal rapid adhesion strengthened by actin cytoskeleton remodeling through Rac and Cdc42. J Cell Biol 167: 1183–1194.

24. .Panorchan P, Thompson MS, Davis KJ, Tseng Y, Konstantopoulos K, et al. (2006) Single-molecule analysis of cadherin-mediated cell-cell adhesion. J of Cell Sc 119: 66–74.

25. Prakasam AK, Maruthamuthu V, Leckband D (2006) Similarities between heterophilic and homophilic cadherin adhesion. Proc Nat Ac Sc USA 103: 15434–15439.

26. Angres B, Barth A, Nelson WJ (1996) Mechanism for transition from initial to stable cell-cell adhesion: kinetic analysis of E-cadherin-mediated adhesion using a quantitative adhesion assay. J Cell Biol 134: 549–557.

27. Adams CL, Nelson WJ (1998) Cytomechanics of cadherin-mediated cell-cell adhesion. Curr Op Cell Biol 10: 572–577.

28. Adams CL, Chen YT, Smith SJ, Nelson WJ (1998) Mechanisms of epithelial cell-cell adhesion and cell compaction revealed by high-

resolution tracking of E-cadheringreen fluorescent protein. J Cell Biol 142: 1105–1119.

29. Behrens J, Birchmeier W, Goodman SL, Imhof BA (1985) Dissociation of Madin-Darby canine kidney epithelial cells by the monoclonal antibody anti-arc-1: mechanistic aspects and identification of the antigen as a component related to uvomorulin. J Cell Biol 101: 1307–1315.

30. Halbleib JM, Nelson WJ (2006) Cadherins in development: cell adhesion, sorting, and tissue morphogenesis. Genes & Development 20: 3199–3214.

31. .Shapiro L, Fannon AM, Kwong PD, Thompson A, Lehmann MS, et al. (1995) Structural basis of cell-cell adhesion by cadherins. Nature 374: 327–337.

32. .Sivasankar S, Brieher W, Lavrik N, Gumbiner BM, Leckband D (1999) Direct molecular force measurements of multiple adhesive interactions between cadherin ectodomains. Proc Nat Ac Sc USA 96: 11820–11824.

33. Chappuis-Flament S, Wong E, Hicks LD, Kay CM, Gumbiner BM (2001) Multiple cadherin extracellular repeats mediate homophilic binding and adhesion. J Cell Biol 154: 231–243.

34. .Zhu B, Chappuis-Flament S, Wong E, Jensen IE, Gumbiner BM, et al. (2003) Functional analysis of the structural basis of homophilic cadherin adhesion. Biophys J 84: 4033–4042.

35. Pertz O, Bozic D, Koch AW, Fauser C, Brancaccio A, et al. (1999) A new crystal structure, Ca^{2+} dependence and mutational analysis reveal molecular details of E-cadherin homoassociation. EMBO J 18: 1738–1747.

36. .Boggon TJ, Murray J, Chappuis-Flament S, Wong E, Gumbiner BM, et al. (2002) C-cadherin ectodomain structure and implications for cell adhesion mechanisms. Science 296: 1308–1313.

37. .Chen CP, Posy S, Ben-Shaul A, Shapiro L, Honig BH (2005) Specificity of cell-cell adhesion by classical cadherins: Critical role for low-affinity dimerization through beta-strand swapping. Proc Nat Ac Sc USA 102: 8531–8536.

38. Graner F, Glazier JA (1992) Simulation of biological cell sorting using a twodimensional extended Potts model. Phys Rev Lett 69: 2013–2016.

39. Glazier JA, Graner F (1993) Simulation of the differential adhesion driven rearrangement of biological cells. Phys Rev E 47: 2128–2154.

40. .Glazier JA, Balter A, Poplawski NJ (2007) pp. 79–106. Single-Cell-Based Models in Biology and Medicine, Birkhauser-Verlag, Basel,

Switzerland, chapter Magnetization to morphogenesis: a brief history of the Glazier-Graner-Hogeweg model.

41. Mombach JC, Glazier JA (1996) Single cell motion in aggregates of embryonic cells. Phys Rev Lett 76: 3032–3035.

42. .Rieu JP, Upadhyaya A, Glazier JA, Ouchi NB, Sawada Y (2000) Diffusion and deformations of single hydra cells in cellular aggregates. Biophys J 79: 1903–1914.

43. .Zajac M, Jones GL, Glazier JA (2000) Model of convergent extension in animal morphogenesis. Phys Rev Lett 85: 2022–2025.

44. .Maree AF, Hogeweg P (2001) How amoeboids self-organize into a fruiting body: multicellular coordination in Dictyostelium discoideum. Proc Nat Ac Sc USA 98: 3879–3883.

45. .Zeng W, Thomas GL, Glazier JA (2004) Non-Turing stripes and spots: a novel mechanism for biological cell clustering. Physica A 341: 482–494.

46. Dan D, Mueller C, Chen K, Glazier JA (2005) Solving the advection-diffusion equations in biological contexts using the cellular Potts model. Phys Rev E 72: 041909.

47. .Merks RHM, Glazier JA (2005) A cell-centered approach to developmental biology. Physica A 352: 113–130.

48. .Poplawski NJ, Swat M, Gens JS, Glazier JA (2007) Adhesion between cells, diffusion of growth factors, and elasticity of the AER produce the paddle shape of the chick limb. Physica A 373C: 521–532.

49. .Glazier JA, Zhang Y, Swat M, Zaitlen B, Schnell S (2008) Coordinated action of N-CAM, N-cadherin, EphA4, and ephrinB2 translates genetic prepattern into structure during somitogenesis in chick. Curr Top Dev Biol 81: 205–247.

50. .Metropolis N, Rosenbluth AW, Rosenbluth MN, Teller AH, Teller E (1953) Equations of State Calculations by Fast Computing Machines. J ChemPhys 21: 1087–1092.

51. .Armstrong PB (1971) Light and electron microscope studies of cell sorting in combinations of chick embryo neural retina and retinal pigment epithelium. Wilhelm Roux' Archiv 168: 125–141.

52. .Frenkel J (1945) Viscous flow of crystalline bodies under the action of surface tension. J Phys 4: 385–431.

53. .Beysens DA, Forgacs G, Glazier JA (2000) Cell sorting is analogous to phase ordering in fluids. Proc Nat Ac Sc USA 97: 9467–9471.

54. Steinhaus H (1956) Sur la division des corps materiels en parties. Bull Acad Polon Sci 4: 801–804.

55.   Watanabe T, Sato Y, Saito D, Tadokoro R, Takahashi Y (2009) EphrinB2 coordinates the formation of a morphological boundary and cell epithelialization during somite segmentation. Proc Natl Acad Sci U S A 106(18): 7467–7472.

Chapter 5

NONEQUILIBRIUM METHODS FOR EQUILIBRIUM FREE ENERGY CALCULATIONS

INTRODUCTION AND BACKGROUND

In this chapter, we will show how nonequilibrium methods can be used to calculate equilibrium free energies. This may appear contradictory at first glance. However, as was shown by Jarzynski [1, 2], nonequilibrium perturbations can be used to obtain equilibrium free energies in a formally exact way. Moreover, Jarzynski's identity also provides the basis for a quantitative analysis of experiments involving the mechanical manipulation of single molecules using, e.g., force microscopes or laser tweezers [3–6].

But before proceeding to nonequilibrium averages, we briefly review two closely related and previously introduced methods for free energy calculations, thermodynamic integration and free energy perturbation theory. In thermodynamic integration, an order parameter is used to describe the transition between two states, and equilibrium averages are used to evaluate derivatives of the free energy with respect to that order parameter. One then integrates the free energy derivatives along a continuous path connecting the initial and final states to obtain the free energy difference between them. As an example, Born [7] chose the charge of the ion as a coupling parameter to estimate the ion-solvation free energy, with the average solvent-induced electrostatic potential at the ion site as the corresponding free energy derivative. In free energy perturbation theory, one avoids (at least in principle) intermediate equilibrium averages and instead uses only an equilibrium ensemble of configurations in the initial state. Each of these configurations is converted instantaneously (i.e., without relaxation of the atom positions, etc.) to the final state and the resulting difference in energy is evaluated. A Boltzmann average of that energy difference then yields the free energy difference between initial and final states.

These two seemingly distinct approaches of thermodynamic integration and perturbation can be seen as the limiting cases of a more general formalism in which the transformation between the two states proceeds at a finite rate. Seen in this light, one might also hope to obtain free energies from a transformation that converts the initial to the final state neither infinitely slowly (as in thermodynamic integration) nor infinitely fast (as in free energy perturbation theory). Rigorous free energy calculations using transformations at a finite rate are indeed possible through Jarzynski's identity [1]. As it turns out, the simulation algorithm implementing the necessary calculations had already been used extensively before, but the simulation data were analyzed in only an approximate manner and under the assumption of near-equilibrium conditions [8–11].

In the following paragraph, free energy perturbation theory and thermodynamic integration will be briefly described to introduce the necessary notation and provide a framework for the subsequent outline of nonequilibrium free energy calculations. We consider a system with phase space coordinates z (i.e., positions and momenta of all atoms) and a Hamiltonian $\mathcal{H}$ (z; λ) that depends parametrically on a coupling parameter λ. When charging a Born ion [7], for instance, λ would be proportional to the charge of the ion, $q = \lambda e$. The Helmholtz free energy A is then a function of λ. In the canonical ensemble, it is given by

$$A(\lambda) = -\beta^{-1} \ln \int \exp[-\beta \mathcal{H}(\mathbf{z}; \lambda)] d\mathbf{z}$$

(5.1)

In the canonical partition function of (5.1), we have for simplicity ignored combinatorial prefactors. Free energy perturbation theory [12] relies on evaluating effectively the ratio of the partition functions, to obtain the free energy difference between the initial and final states corresponding to coupling parameters $\lambda = 1$ and 0,

$$\exp\{-\beta[A(1) - A(0)]\} = \frac{\int \exp[-\beta \mathcal{H}(\mathbf{z}; 1)] d\mathbf{z}}{\int \exp[-\beta \mathcal{H}(\mathbf{z}; 0)] d\mathbf{z}}$$
$$= \langle \exp\{-\beta[\mathcal{H}(\mathbf{z}; 1) - \mathcal{H}(\mathbf{z}; 0)]\} \rangle_0$$
$$= \langle \exp\{+\beta[\mathcal{H}(\mathbf{z}; 1) - \mathcal{H}(\mathbf{z}; 0)]\} \rangle_1^{-1},$$

(5.2)

where the forward and backward averages $\langle \cdots \rangle_0$ and $\langle \cdots \rangle_1$ are evaluated in the reference states $\lambda = 0$ and 1, respectively. Free energy differences can also be obtained from thermodynamic integration by calculating derivatives of the free energy with respect to the coupling parameter. The derivative can be expressed as an ensemble average

$$\frac{\partial A(\lambda)}{\partial \lambda} = \frac{\int \frac{\partial \mathcal{H}}{\partial \lambda} \exp[-\beta \mathcal{H}(\mathbf{z}; \lambda)] d\mathbf{z}}{\int \exp[-\beta \mathcal{H}(\mathbf{z}; \lambda)] d\mathbf{z}}$$

$$= \left\langle \frac{\partial \mathcal{H}}{\partial \lambda} \right\rangle_\lambda \tag{5.3}$$

where the average $\langle \cdots \rangle_\lambda$ is in state λ. To obtain the difference in free energy between states defined by coupling parameter values of 0 and 1, one can then integrate (5.3)

$$A(1) - A(0) \equiv \Delta A = \int_0^1 \left\langle \frac{\partial \mathcal{H}}{\partial \lambda} \right\rangle_\lambda d\lambda . \tag{5.4}$$

Without explicit analytical expressions for the free energy derivative (as could be obtained for the case of a Born ion in a dielectric medium), the integral has to be evaluated numerically by simulation.

As an alternative approach to thermodynamic integration, we now consider creating an equilibrium configuration in state $\lambda = 0$ and then slowly changing λ from 0 to 1 [8, 9]. As the coupling parameter is advanced, the system continues to sample phase space (e.g., by molecular dynamics or Monte Carlo simulations), but under an explicitly time-dependent Hamiltonian (because of the evolving coupling parameter). In the limit of a very slow transformation, with some caveats for Hamiltonian dynamics [13], the system will remain close to equilibrium. The integral in (5.4) can then be evaluated by changing λ continuously

$$\Delta A = \lim_{\tau \to \infty} \int_0^\tau \left. \frac{\partial \mathcal{H}}{\partial \lambda} \right|_{\lambda = \lambda(t)} \dot{\lambda}(t) dt , \tag{5.5}$$

where the time derivative of the coupling parameter λ is denoted by a dot. The coupling parameter λ moves along a continuous prescribed path $\lambda = \lambda(t)$ connecting the initial [$\lambda(0) = 0$] to the final state [$\lambda(\tau) = 1$]. The time interval of the transformation measures the actual time (in molecular dynamics) or the number of steps (e.g., in Monte Carlo simulations). In (5.5), the limit of $\tau \to \infty$ ensures that the transformation is performed infinitely slowly, and thus reversibly. The right-hand side of (5.4) correspondingly defines the "reversible work" done on the system during the transformation.

If the system is instead transformed between initial and final states over a finite time interval τ, the free energy obtained from (5.5) without the $\tau \to \infty$ limit will only be approximate. In the corresponding "slow-growth" method of free energy calculations [8, 9], the system will not be able to sample the phase space exhaustively at each value of λ, rendering the transformation irreversible. As the transformation proceeds, the system will be gradually driven out of equilibrium, causing hysteresis effects. From the second law of thermodynamics, we expect that the work W(τ) performed during the nonequilibrium transformation is on average larger than or equal to the free energy difference between the two states

$$\langle W(\tau) \rangle \geq \Delta A .$$

(5.6)

with the difference between the work and free energy accounting for heat dissipation effects. The work W(τ) performed on the system is the accumulated energetic cost (or gain) required to change the system

$$W(\tau) = \int_0^\tau \frac{\partial \mathcal{H}[\mathbf{z}(t); \lambda]}{\partial \lambda}\bigg|_{\lambda=\lambda(t)} \dot{\lambda}(t) \mathrm{d}t .$$

(5.7)

The average $\langle \cdots \rangle$ in (5.6) is over many repetitions of the transformation initiated from an equilibrium ensemble of configurations at λ = 0, with the same prescribedtime evolution of the coupling parameter λ(t). The equality in (5.6) will normally be achieved only if the transformation is infinitely slow, τ → ∞. Otherwise, the work performed during the transformation will, on average, exceed the free energy difference between the two states. For paths of finite length, the amount of dissipated work, <W(τ)> − ΔA ≥ 0, will depend on the chosen transformation path λ(t).

In the following, we first derive Jarzynski's identity by exploiting its close relation to the Feynman–Kac identity of path integrals (Sects. 5.2 and 5.3). In addition, we present explicit derivations for Hamiltonian dynamics, and illustrate the theory for the analytically solvable case of a moving harmonic oscillator. We then introduce and derive the Crooks identity (Sects. 5.4 and 5.5) in which the perturbation is performed in both forward and backward directions. Section 5.6 covers the implementation of nonequilibrium free energy estimates in computer simulations, including time integration, choice of order parameters, creation of initial conditions, and allocation of computer time. The analysis of the simulation results and the choice of free energy estimators are discussed in Sect. 5.7. A simple nonequilibrium perturbation is used for illustration (Sect. 5.8). Section 5.9 describes how nonequilibrium methods such as steered molecular dynamics simulations can be used to determine potentials of mean force. The chapter concludes with a discussion of selected applications, and a summary

Jarzynski's Identity

Jarzynski has shown that even for nonequilibrium paths, the inequality (5.6) can be turned into an equality [1]. Jarzynski's identity states that

$$\langle \exp[-\beta W(\tau)] \rangle = \exp(-\beta \Delta A) \tag{5.8}$$

where a prescribed path $\lambda(t)$ connects the initial and final states, with $\lambda(0) = 0$ and $\lambda(\tau) = 1$, and $\Delta A = A(1) - A(0)$ the free energy difference between the two states. The average $\langle \cdots \rangle$ in (5.8) requires some explanation. It is a combination of an ensemble average over initial conditions, and a path average over trajectory realizations. Initial conditions are chosen according to the equilibrium Boltzmann probability in the $\lambda(0)$ state. The path average samples all realizations of dynamic paths, weighted by their respective path action, under the time evolution of the system with an explicitly time-varying Hamiltonian. For deterministic dynamics, only a single trajectory exists for any given initial condition, making the path average unnecessary. For stochastic dynamics, the path average is over realizations of noise. If the stochastic trajectories are generated using standard integrators, then the trajectory average is over different sequences of uncorrelated random numbers.

Jarzynski's identity, (5.8), immediately leads to the second law in the form of (5.6) because of Jensen's inequality, $\langle e^{-x} \rangle \geq e^{-\langle x \rangle}$. Moreover, in the limit of an infinitely fast transformation, $\tau \to 0$, we recover free energy perturbation theory. In that limit, the configurations will not relax during the transformation. The average in (5.8) is then only over the initial conditions, and the work is simply the change in the Hamiltonian when going from initial to final state,

$$\lim_{\tau \to 0} W(\tau) = \mathscr{H}(\mathbf{z}(0); \lambda = 1) - \mathscr{H}(\mathbf{z}(0); \lambda = 0) , \tag{5.9}$$

making (5.2) a limiting case of (5.8). In the opposite limit of $\tau \to \infty$, i.e., an infinitely slow transformation, the system is changed reversibly, and we recover thermodynamic integration in the form of (5.5). In the following, a brief "derivation" of Jarzynski's identity will be given.

Derivation of Jarzynski's Identity

As shown in the following, Jarzynski's identity follows almost immediately [3] from the Feynman–Kac theorem for path integrals [14]. A brief and more "pedagogical" discussion of the Feynman–Kac identity in the context of nonequilibrium free energy calculations as well as chemical kinetics, line-shape theory, and quantum mechanics can be found in [15]. The derivation of [3] also leads to an extension of Jarzynski's identity, which will allow us to calculate potentials of mean force for fluctuating observables, in addition to free energy differences between states described by control parameters, i.e., coupling parameters under external control. A free energy relation for fluctuating variables (i.e., variables that are not under external control, such as the instantaneous distance between two particles) proved central for the analysis of certain nonequilibrium experiments and computer simulations. Following [3], we assume that the phase space density $\rho(z,t)$ of the system evolves according to a Liouville-type equation:

$$\frac{\partial \rho(\mathbf{z}, t)}{\partial t} = L_t \rho(\mathbf{z}, t),$$

(5.10)

where L_t is an explicitly time-dependent evolution operator. The time evolution governed by the Liouville operator L_t is consistent with the Hamiltonian $\mathscr{H}$ (z,t). As an example, L_t for classical dynamics will be given below explicitly in terms of the Hamiltonian $\mathscr{H}$ (5.20), $\mathscr{H}$ depends parametrically on time, thus in effect absorbing the parametric dependences of $\mathscr{H}$ on λ, and of λ on t. As an example, for Smoluchowski diffusion (i.e., Brownian dynamics) on a time-dependent potential energy surface V (z,t), the Liouville operator would be $L_t = D\nabla e^{-\beta V(\mathbf{z},t)} \nabla e^{\beta V(\mathbf{z},t)}$, where D is the diffusion coefficient and $\nabla = \partial/\partial z$.

We assume further that the Boltzmann distribution is stationary, $L_t e^{-\beta \mathscr{H}(\mathbf{z},t)}$=0. This requirement is satisfied, for instance, by systems that undergo Newtonian, Langevin, or Metropolis Monte Carlo dynamics. For diffusion, $\mathscr{H}$ (z,t) would just be the potential energy surface V (z,t), and the corresponding Boltzmann distribution is indeed stationary under the L_t given in the preceding paragraph. To avoid confusion, note that the stationarity of the Boltzmann distribution under the Liouville operator, $L_t e^{-\beta \mathscr{H}(\mathbf{z},t')}= 0$, is in general only valid at equal times t = t', i.e., $L_t e^{-\beta \mathscr{H}(\mathbf{z},t')} \neq 0$ for $t \neq t'$ in most cases. In particular, if one starts at time t = 0

from a Boltzmann distribution corresponding to $\mathscr{H}$ (z, 0) and evolves that distribution under L_t, then at some time t > 0 that distribution will no longer be stationary, $L_t e^{-\beta \mathscr{H}(\mathbf{z},0)} \neq 0$, and the system will be driven out of equilibrium.

Now consider the un-normalized Boltzmann distribution at time t

$$\rho(\mathbf{z}, t) = \frac{e^{-\beta \mathscr{H}(\mathbf{z},t)}}{\int e^{-\beta \mathscr{H}(\mathbf{z}',0)} d\mathbf{z}'}.$$

(5.11)

Since this distribution is stationary with respect to L_t (i.e., $L_t\rho(\mathbf{z},t) = 0$), and since $\partial\rho/\partial t = -\beta(\partial\mathcal{H}/\partial t)\rho$, it follows that the above $\rho(z,t)$ is a solution of the sink (or birth–death) equation

$$\frac{\partial\rho}{\partial t} = L_t\rho - \beta\frac{\partial\mathcal{H}}{\partial t}\rho,$$

(5.12)

as can be verified by direct substitution. If $\partial\mathcal{H}/\partial t \equiv 0$, (5.12) is the Liouville equation. Otherwise, the "sink" term locally adds phase space density where $\partial\mathcal{H}(z,t)/\partial t$ is negative, and removes density where it is positive. Qualitatively, one may think of the addition and subtraction of phase space density as compensating for the fact that trajectories evolving under the time-dependent Liouville operator (Hamiltonian) equilibrate too slowly to maintain the time-dependent Boltzmann distribution, (5.11).

The solution of the sink equation (5.12), starting from an equilibrium distribution at time t = 0, can also be expressed as a path integral by using the Feynman–Kac theorem,

$$\rho(\mathbf{z},t) = \left\langle \delta(\mathbf{z} - \mathbf{z}(t))e^{-\beta\int_0^t \frac{\partial\mathcal{H}}{\partial t'}(\mathbf{z}(t'),t')dt'} \right\rangle$$

(5.13)

where $\delta(x)$ is Dirac's delta distribution. A motivation of the Feynman–Kac theorem [14] can be found in [15]. In essence, the Feynman–Kac theorem provides a way to reweight trajectories generated without consideration of the sink term in (5.12) to take the sink term into account. In the path average of (5.13), trajectories are generated according to the Liouville operator L_t, without consideration of the sink (or birth-death) term $-\beta(\partial\mathcal{H}/\partial t)\rho$, starting from an initial distribution $\rho(z, 0)$. The sink term is taken into account by weighting trajectory end points z(t). Consider S(t) the relative weight of a trajectory end point at time t. Along a given trajectory, that weight will change during the time interval $(t_t + dt)$ by $dS = -\beta(\partial\mathcal{H}/\partial t)S(t)dt$ according to the sink-term in (5.13). With S(0) = 1, the aggregate weight accumulated along the trajectory is then $S(t) = \exp[-\beta\int_0^t (\partial\mathcal{H}/\partial t')dt']$, the weight factor in (5.13).

Equating the two different solutions of (5.12), (5.11), and (5.13) immediately gives

$$\frac{e^{-\beta \mathcal{H}(\mathbf{z},t)}}{\int e^{-\beta \mathcal{H}(\mathbf{z}',0)} d\mathbf{z}'} = \left\langle \delta(\mathbf{z} - \mathbf{z}(t)) e^{-\beta \int_0^t \frac{\partial \mathcal{H}}{\partial t'}(\mathbf{z}(t'), t') dt'} \right\rangle$$

(5.14)

This identity [3, 15] between a weighted average of nonequilibrium trajectories (r.h.s.) and the equilibrium Boltzmann distribution (l.h.s.) is implicit in the work of Jarzynski [2], and is given explicitly by Crooks [16]. The average $\langle \cdots \rangle$ is over an ensemble of trajectories starting from the equilibrium distribution at t = 0 and evolving according to (5.10). Each trajectory is weighted with the Boltzmann-factor of the external work W(t) done on the system,

$$W(t) = \int_0^t \frac{\partial \mathcal{H}(\mathbf{z}(t'), t')}{\partial t'} dt' .$$

(5.15)

By integrating both sides of (5.14) with respect to z, we obtain Jarzynski's identity [1, 2]

$$e^{-\beta \Delta A(t)} \equiv \frac{\int e^{-\beta \mathcal{H}(\mathbf{z},t)} d\mathbf{z}}{\int e^{-\beta \mathcal{H}(\mathbf{z},0)} d\mathbf{z}} = \left\langle e^{-\beta W(t)} \right\rangle ,$$

(5.16)

between the Boltzmann-averaged work W(t) and the equilibrium free energy difference $\Delta A(t)$ between times t and 0.

The above derivation shows that Jarzynski's identity is an immediate consequence of the Feynman–Kac theorem. This connection has not only theoretical value, but is also useful in practice. First, it forms the basis for an equilibrium thermodynamic analysis of nonequilibrium pulling experiments [3, 15]. Second, it helps in deriving a Jarzynski identity for dynamics using thermostats. Moreover, this derivation clari- fies an important aspect: trajectories can be thought of as mapping initial conditions (t = 0) to trajectory endpoints, and the Boltzmann factor of the accumulated work reweights that map to give the desired Boltzmann distribution. Finally, it can be easily extended to transformations between steady states [17] in which non-Boltzmann distributions are stationary.

Hamiltonian Dynamics

In the following, we will show explicitly that the identity (5.8) holds for deterministic Newtonian dynamics. This derivation builds on the arguments given by Jarzynski in his original paper [1]. The essence of the derivation is: (1) integration of a trajectory for a fixed time t uniquely maps an initial phase point $z(0)$ to a final phase point $z(t)$; (2) an initial phase space density remains unchanged along trajectories because of Liouville's theorem: $\rho(z(t),t)$ = $\rho(z(0), 0)$; (3) the work along the trajectory is exactly the change in energy, $W(t) = \mathcal{H}(z(t),t) - \mathcal{H}(z(0), 0)$; such that (4) weighting the trajectory endpoints by $\exp(-\beta W(t))$ transforms an initial phase space distribution $\rho(z(0), 0)$ $\propto \exp(-\beta \mathcal{H}(z(0), 0))$ to a distribution proportional to $\exp(-\beta \mathcal{H}(z(t),t))$, i.e., a Boltzmann distribution for the final Hamiltonian.

Let us consider a system described by an explicitly time-dependent Hamiltonian $\mathcal{H}(p, q,t)$ where $(p, q) = z$ is a point in phase space. Hamilton's equation of motion are

$$\dot{\mathbf{p}} = -\frac{\partial \mathcal{H}}{\partial \mathbf{q}} \; ; \quad \dot{\mathbf{q}} = \frac{\partial \mathcal{H}}{\partial \mathbf{p}} \; ; \quad \frac{d\mathcal{H}}{dt} = \frac{\partial \mathcal{H}}{\partial t}$$

(5.17)

The probability density $\rho(p, q,t)$ in phase space satisfies the continuity equation

$$\frac{\partial \rho}{\partial t} + \frac{\partial}{\partial \mathbf{p}}(\rho \dot{\mathbf{p}}) + \frac{\partial}{\partial \mathbf{q}}(\rho \dot{\mathbf{q}}) = 0$$

(5.18)

From Hamilton's equations, we find that $\partial \dot{\mathbf{p}}/\partial \mathbf{p} = -\partial \dot{\mathbf{q}}/\partial \mathbf{q}$, such that

$$\frac{\partial \rho}{\partial t} + \dot{\mathbf{p}}\frac{\partial \rho}{\partial \mathbf{p}} + \dot{\mathbf{q}}\frac{\partial \rho}{\partial \mathbf{q}} = 0 .$$

(5.19)

This is Liouville's equation, with the Liouville operator

$$L_t = \dot{\mathbf{p}}(\partial/\partial \mathbf{p}) + \dot{\mathbf{q}}(\partial/\partial \mathbf{q}) = (\partial \mathcal{H}/\partial \mathbf{p})(\partial/\partial \mathbf{q}) - (\partial \mathcal{H}/\partial \mathbf{q})(\partial/\partial \mathbf{p})$$ (5.20)

Equation (5.19) states that the flow in phase space is incompressible. In particular, following along a trajectory starting in $(p(0), q(0))$, we find that

$$\frac{d\rho(\mathbf{p}(t), \mathbf{q}(t), t)}{dt} = \frac{\partial \rho}{\partial t} + \dot{\mathbf{p}}\frac{\partial \rho}{\partial \mathbf{p}} + \dot{\mathbf{q}}\frac{\partial \rho}{\partial \mathbf{q}} = 0 ,$$

(5.21)

where we used the chain rule with respect to p and q, and (5.19). This means that the phase space density along a trajectory remains unchanged

$$\rho(\mathbf{p}(t), \mathbf{q}(t), t) = \rho(\mathbf{p}(0), \mathbf{q}(0), 0) .$$

(5.22)

From the definition of the work, (5.15), and Hamilton's equations, (5.17), it follows that the work performed along the trajectory is

$$W(t) = \int_0^t \frac{\partial \mathcal{H}(\mathbf{p}(t'), \mathbf{q}(t'), t')}{\partial t'} dt' = \int_0^t \frac{d\mathcal{H}(\mathbf{p}(t'), \mathbf{q}(t'), t')}{dt'} dt'$$

$$= \mathcal{H}(\mathbf{p}(t), \mathbf{q}(t), t) - \mathcal{H}(\mathbf{p}(0), \mathbf{q}(0), 0) .$$

(5.23)

If we choose our initial density according to a Boltzmann distribution, ρ(p(0), q(0), 0) = exp(−β$\mathcal{H}$ (p(0), q(0), 0))/ = dp dq exp(−β$\mathcal{H}$ (p, q, 0)), the density at the trajectory end point at time t is identical to that at the initial point, ρ(p(t), q(t),t) = exp(−β$\mathcal{H}$ (p(0), q(0), 0))/ = dp dq exp(−β$\mathcal{H}$ (p, q, 0)). If we weight each trajectory end point by the Boltzmann factor of the accumulated work, exp(−βW(t)), and use (5.23) for W(t), we obtain a "Jarzynski-modified" density ρJ given by

$$\rho_J(\mathbf{p}(t), \mathbf{q}(t), t) = \rho(\mathbf{p}(t), \mathbf{q}(t), t) \exp[-\beta W(t)]$$

$$= \frac{\exp[-\beta \mathcal{H}(\mathbf{p}(t), \mathbf{q}(t), t)]}{\int d\mathbf{p} \, d\mathbf{q} \exp[-\beta \mathcal{H}(\mathbf{p}, \mathbf{q}, 0)]} .$$

(5.24)

That is, we have recovered a Boltzmann distribution according to the Hamiltonian at time t, equivalent to (5.14). Jarzynski's identity (5.8) then follows simply by integration over phase space (p, q).

This derivation addresses a potentially confusing point, namely that the final state at time t is not at equilibrium and may not have a well-defined temperature. As is clear from the derivation, temperature here is only a parameter that is once used to specify the initial condition, and then again in the Boltzmann weight of the integrated work. From this viewpoint, the reweighting is a convenient mathematical "trick."

Moving Harmonic Oscillator

To further illustrate the theory, we apply Jarzynski's identity to the analytically solvable example of a 1D moving harmonic oscillator with Hamiltonian

$$\mathcal{H}(p, q, t) = \frac{p^2}{2m} + \frac{k}{2}(q - vt)^2$$

(5.25)

where m is the mass, k is the spring constant, and v is the pulling velocity. Because the time dependence of the Hamiltonian simply induces a translational shift, the free energy remains unchanged

$$A(t) = -\beta^{-1} \int dp \int dq \; e^{-\beta \mathscr{H}(p,q,t)} \equiv A(0)$$

(5.26)

In the following, we will show explicitly that the correct result is obtained if Jarzynski's identity is used to evaluate the free energy difference, $A(t) - A(0)$ = $\langle \exp(-\beta W) \rangle$ = 0.

For given initial conditions, trajectories in phase space satisfying Hamilton's equation of motion, (5.17), are given by

$$\begin{pmatrix} p(t) \\ q(t) \end{pmatrix} = \begin{pmatrix} \cos(\alpha t) & -(km)^{1/2}\sin(\alpha t) \\ (km)^{-1/2}\sin(\alpha t) & \cos(\alpha t) \end{pmatrix} \cdot \begin{pmatrix} p(0) \\ q(0) \end{pmatrix}$$
$$+ \begin{pmatrix} 2mv\sin^2\left(\frac{\alpha t}{2}\right) \\ vt - \frac{vm^{1/2}}{k^{1/2}}\sin(\alpha t) \end{pmatrix},$$

(5.27)

where $\alpha = (k/m)^{1/2}$. As required by the Liouville theorem (5.22), the map of initial conditions $(p(0),q(0))$ to phase points $(p(t),q(t))$ at time t has a Jacobian equal to one, $|\partial(p(t),q(t))/\partial(p(0),q(0))| = 1$. According to (5.15) and (5.23), the integrated work along a trajectory is given by

$$W(t) = \int_0^t \frac{\partial \mathscr{H}}{\partial t'} dt' = v[mv - p(0)]\left[1 - \cos\left(\frac{k^{1/2}t}{m^{1/2}}\right)\right]$$
$$-v(km)^{1/2}q(0)\sin\left(\frac{k^{1/2}}{m^{1/2}}t\right)$$

(5.28)

which is a linear function with respect to the initial phase space coordinates $(p(0), q(0))$. Note that so far, temperature has not appeared!

Now we specify an initial phase space distribution as the Boltzmann distribution for the Hamiltonian at time $t = 0$

$$\rho(q = q(0), p = p(0), t = 0) = \beta \frac{\exp\left(-\frac{\beta p^2}{2m} - \frac{\beta k q^2}{2}\right)}{2\pi(m/k)^{1/2}}$$

(5.29)

With this distribution being Gaussian in both $p(0)$ and $q(0)$, and the work being linear in $p(0)$ and $q(0)$, the distribution of the work obtained for Boltzmann-weighted initial conditions is Gaussian as well, with a mean and variance of

$$\langle W \rangle = 2mv^2 \sin^2\left(\frac{k^{1/2}t}{2m^{1/2}}\right)$$

(5.30)

$$\sigma^2(W) = 2\beta^{-1} \langle W \rangle$$

$$(5.31)$$

For a Gaussian distribution of the work, second-order perturbation theory is exact, and $\langle \exp(-\beta W) \rangle = \exp[-\beta \langle W \rangle + \beta^2 \sigma^2(W)/2]$. We thus find that the free energy difference for the moving oscillator according to (5.8) is $A(t) - A(0) = \langle W \rangle - \beta \sigma^2(W)/2 \equiv 0$, which is indeed the desired result (expected from translation invariance).

We note that for the harmonic Hamiltonian in (5.25), the variance of the work approaches zero in the limit of an infinitely slow transformation, $v \to 0$, $\tau \to \infty$, $v\tau = $ const. However, as shown by Oberhofer et al. [13], this is not the case in general. As a consequence of adiabatic invariants of Hamiltonian dynamics, even in- finitely slow transformations can result in a non-delta-like distribution of the work. Analytically solvable examples for that unexpected behavior are, for instance, harmonic Hamiltonians with time-dependent spring constants $k = k(t)$.

Forward and Backward Averages: Crooks Relation

So far, we have only considered perturbations in the forward direction. As in conventional free energy calculations, powerful relations can be derived if forward and backward perturbations are combined. With the free energy being a state function, we can reverse the path direction. This leads to several useful relations derived originally by Crooks [16, 18, 19]. In particular, we obtain immediately

$$\langle \exp[-\beta \underline{W}(\tau)] \rangle = \exp\{-\beta[A(0) - A(1)]\} \equiv \exp(\beta \Delta A),$$

$$(5.32)$$

where $\underline{W}(\tau)$ is the work accumulated on the reversed path, $\underline{\lambda}(t) = \lambda(\tau - t)$

$$\underline{W}(\tau) = \int_0^\tau \frac{\partial \mathscr{H}[\mathbf{z}(t); \lambda]}{\partial \lambda}\bigg|_{\lambda=\underline{\lambda}(t)} \dot{\underline{\lambda}}(t)\mathrm{d}t.$$

$$(5.33)$$

In combination with (5.6), this leads to an upper and lower bound for the free energy difference

$$-\langle \underline{W}(\tau) \rangle \leq \Delta A \leq \langle W(\tau) \rangle .$$

$$(5.34)$$

These bounds are the nonequilibrium equivalents of the Gibbs–Bogoliubov bounds discussed. Having the free energy now bounded from above and below already demonstrates the power of using both forward and backward transformations. Moreover, as was shown by Crooks [18, 19], the distribution of work values from forward and backward paths satisfies a relation that is central to histogram methods in free energy calculations

$$\frac{p_f[w = W(\tau)]}{p_b[w = -\underline{W}(\tau)]} = \exp[\beta(w - \Delta A)]$$

(5.35)

where $p_f[w = W(\tau)]$ and $p_b[w = -\underline{W}(\tau)]$ are the probability densities of the work values for forward and reversed transformation paths (with a sign change in the work of the reverse path). Both are normalized, i.e., $\int p_f(w)dw = \int p_f(w)dw = 1$.

For computer simulations, (5.35) leads to accurate estimates of free energies. It is also the basis for higher-order cumulant expansions [20] and applications of Bennett's optimal estimator [21–23]. We note that Jarzynski's identity (5.8) follows from (5.35) simply by integration over w because the probability densities are normalized to 1:

$$\int p_f(W)e^{-\beta W}dW = \int e^{-\beta \Delta A}p_b(W)dW$$

(5.36)

Because of the normalization condition, the right-hand side is equal to $\exp(-\beta\Delta A)$, and Jarzynski's identity follows.

Derivation of the Crooks Relation (and Jarzynski's Identity)

The Crooks relation follows from an elegant derivation of Jarzynski's identity using path-sampling ideas [18]. For instructive purposes, that derivation is briefly summarized here. Consider generating a discrete trajectory $z_0 \xrightarrow{\mathcal{H}_1} z_1 \xrightarrow{\mathcal{H}_2} \ldots \xrightarrow{\mathcal{H}_N} z_N$ where each step $z_{i-1} \xrightarrow{\mathcal{H}_i} z_i$ proceeds according to Hamiltonian $\mathcal{H}_i$ for instance by using Metropolis Monte Carlo, Newtonian, or Langevin dynamics "integrators." After each "time" step, the Hamiltonian is changed in a prescribed way. If each of the steps is Markovian in the full phase space (i.e., it does not depend explicitly on the preceding path), then the probability Pf of generating a particular trajectory can be factorized

$$P_f(z_0 \xrightarrow{\mathcal{H}_1} z_1 \xrightarrow{\mathcal{H}_2} \ldots \xrightarrow{\mathcal{H}_N} z_N) = p_0(z_0) \prod_{i=1}^{N} p_i(z_i|z_{i-1}),$$

(5.37)

where $p_0(z_0) = \exp[-\beta(\mathcal{H}_0(z_0) - A_0)]$ is the normalized equilibrium Boltzmann probability density at the initial state (with $A_0 = A(0)$ its free energy), and $p_i(z_i|z_{i-1})$ is the transition probability from z_{i-1} to z_i under the influence of Hamiltonian $\mathcal{H}_i$. Now if these transition probabilities satisfy detailed balance (as they do under Newtonian, Langevin, or Metropolis Monte Carlo dynamics)

$$\frac{p_i(\mathbf{z}_i|\mathbf{z}_{i-1})}{p_i(\mathbf{z}_{i+1}|\mathbf{z}_i)} = e^{-\beta[\mathcal{H}_i(\mathbf{z}_i)-\mathcal{H}_i(\mathbf{z}_{i-1})]}$$

(5.38)

then the probability P_b of picking the time-reversed path under the influence of the corresponding time-dependent Hamiltonian can be related to the probability P_f of the forward path:

$$\frac{P_f}{P_b} = \frac{P(\mathbf{z}_0 \overset{\mathcal{H}_1}{\rightarrow} \mathbf{z}_1 \overset{\mathcal{H}_2}{\rightarrow} \ldots \overset{\mathcal{H}_N}{\rightarrow} \mathbf{z}_N)}{P(\mathbf{z}_N \overset{\mathcal{H}_N}{\rightarrow} \mathbf{z}_{N-1} \overset{\mathcal{H}_{N-1}}{\rightarrow} \ldots \overset{\mathcal{H}_1}{\rightarrow} \mathbf{z}_0)} = \frac{p_0(\mathbf{z}_0) \prod_{i=1}^{N} p_i(\mathbf{z}_i|\mathbf{z}_{i-1})}{p_N(\mathbf{z}_N) \prod_{i=1}^{N} p_i(\mathbf{z}_{i-1}|\mathbf{z}_i)}$$

(5.39)

If we now use (5.38) and substitute the equilibrium Boltzmann distributions for $p_0(\mathbf{z}_0)$ and $p_N(\mathbf{z}_N)$, we obtain

$$\frac{P_f}{P_b} = \exp\left[\beta \sum_{i=0}^{N-1} [\mathcal{H}_{i+1}(\mathbf{z}_i) - \mathcal{H}_i(\mathbf{z}_i)] - \beta(A_N - A_0)\right] = e^{\beta(W-\Delta A)}$$

(5.40)

where we used that the work W is the accumulated change in the energy

$$W = \sum_{i=0}^{N-1} [\mathcal{H}_{i+1}(\mathbf{z}_i) - \mathcal{H}_i(\mathbf{z}_i)] .$$

(5.41)

A_N and A_0 are the free energies corresponding to Hamiltonians $\mathcal{H}_N$ and $\mathcal{H}_0$, with $\Delta A = A_N - A_0$. Equation (5.40) generalizes detailed balance to nonequilibrium trajectories. For every forward path, there is an equivalent backward path whose path probability differs exactly by a factor $\exp[\beta(W - \Delta A)]$. By construction, the work accumulated on that backward path is $\underline{W} = -W$. The distribution of the work is obtained by sampling all paths with their appropriate weights. In particular, by virtue of (5.40), we can obtain the distribution of work on forward paths also by sampling reweighted backward paths. Because the reweighting factor depends only on the work (and on ΔA), we can apply it to the work distribution directly to obtain the Crooks relation, (5.35). We note that for "reversible" paths, the work distributions will be delta functions, $p_f(W) = p_b(W) = \delta(W - \Delta A)$, and the forward and backward paths have the same path probability.

Implementation

Dynamics

Implementing Jarzynski's identity in a free energy calculation is relatively straightforward. Many simulation packages already contain code to perform slow-growth thermodynamic integration. Two considerations are important. First, one is now integrating Newton's equations of motion for a time-dependent Hamiltonian, such that the errors in the trajectories z(t) can be larger. Integration errors are particularly relevant for fast transformations. Second, one needs to evaluate the work along a trajectory. This will normally require an approximation to the time integral, (5.15), such as a trapezoidal rule

$$W(\tau) \approx \sum_{i=0}^{N-1} \frac{t_{i+1} - t_i}{2} \left(\left. \frac{\partial \mathscr{H}}{\partial t} \right|_{t=t_i} + \left. \frac{\partial \mathscr{H}}{\partial t} \right|_{t=t_{i+1}} \right)$$

with $t_0 = 0$ and $t_N = \tau$. For Hamiltonian dynamics, one could also use (5.23) and evaluate the total energy at the beginning and end of the trajectory.

In tests using the moving 1D Hamiltonian harmonic oscillator, (5.25), a velocity-Verlet integrator [24] combined with trapezoidal integration of W(t) performed well when compared to the analytic solution. An interesting analysis of how long time steps can be used to accelerate the calculation without sacrificing accuracy can be found in [25].

Alternatively, one can break up the trajectory dynamics and the change in the coupling parameter into two substeps. In the first half-step, the phase point is advanced by using the short-time expansions of the propagators of the corresponding Liouville or Fokker–Planck equation [24]. In the second half step, the coupling parameter is changed at a fixed configuration, and the corresponding work accumulated,

$$W(t + \Delta t) = W(t) + \mathscr{H}[\mathbf{z}(t), \lambda(t + \Delta t)] - \mathscr{H}[\mathbf{z}(t), \lambda(t)] \quad (5.42)$$

This procedure follows, in effect, the derivation of Jarzynski's identity in discrete time [2, 18], as outlined in Sect. 5.5. Finally, for Hamiltonian dynamics, one can use (5.28) and calculate the work directly from the difference in total energy between trajectory start and end point.

Choice of Coupling Parameter

A critical element for the success of the calculations is the choice of a suitable coupling parameter λ and a path $\lambda(t)$ that connects the initial and final states. If the path is poorly chosen, the system will be driven out of equilibrium quickly and the amount of dissipated work, $\langle W(\tau) \rangle - \Delta A \geq 0$, will be large. As a result, the free energy estimate will likely be inaccurate. Finding good transformation paths may require some intuition about how one can connect states smoothly, i.e., without encountering high energy barriers on typical paths. The same general rules apply as in other free energy calculations. Specifically in the context of nonequilibrium free energy calculations, Reinhardt and coworkers [26, 27] have used variational optimization to minimize the amount of dissipated work. A seemingly different approach would be to determine a good (kinetically relevant) "reaction coordinate" [28], and use that to define $\lambda(t)$.

Creation of Initial Conditions

Before trajectories can be started, an ensemble of initial conditions has to be created. Having a good representation of relevant conformations in the initial state will be particularly important if the transformation is performed relatively fast. In contrast, if the transformation is performed slowly, the system will sample relevant phase space regions for $\lambda \approx 0$ as it slowly moves away from the initial state. Ensembles of equilibrium initial conditions can be taken from a sufficiently long equilibrium simulation, possibly with accelerated sampling using methods discussed in other chapters. An important consideration is that the initial configurations are sufficiently uncorrelated. Otherwise, one may expect correlations among the resulting work values that interfere with applications of "optimal" free energy estimators [21, 22]. A similar issue of creating representative initial conformations arises, for instance, in the calculation of rate coefficients where equilibrium configurations along the transition state surface have to be generated [29, 30]. One simple way of extending an ensemble of configurations is by resampling velocities from a Maxwell–Boltzmann distribution.

Allocation of Computer Time

The allocation of computer time requires some care. In particular, one must decide whether to use many short simulation runs ("fast growth") or few long runs ("slow-growth"). In general, one expects the latter to give more accurate results for a given amount of computer time [20]. As a rule of thumb, the runs should be slow enough that the standard deviation in the resulting work values is about k_BT. If the standard deviation is smaller, the statistical error in the free energy estimate is only insignificantly smaller than that from a single long run, but from multiple runs one also obtains an estimate of that error. If the standard deviation of the work is much larger, in either the forward or backward direction [31], sampling errors will arise because the estimate of $\langle \exp(-\beta W) \rangle$ will be dominated by only one or a few trajectories with low W values, as discussed in the following. Of relevance in this context is a recent result by Jarzynski [31] who showed that among forward and backward transformations, the process with the wider work distribution provides a better estimate of the free energy (just as test particle insertion usually gives better estimates than particle removal in calculations of chemical potentials [32]).

Analysis of Nonequilibrium Free Energy Calculations

Exponential Estimator – Issues with Sampling Error and Bias

One of the challenges in using Jarzynski's identity for free energy calculations is the exponential weight $\exp(-\beta W(t))$ given to trajectories. If the work has a broad distribution with respect to $k_B T$, only a few trajectories at the lower tail of the work distribution will contribute significantly to the weighted average, with all others having "exponentially" small relative weights. In more physical terms, when the transformation is conducted rapidly, most trajectories do not sample relevant regions of phase space. Rapid crossing of barriers under tension, and crossing over atypically high barriers result in hysteresis effects, and a broadened work distribution that reflects an increasing relevance of dissipation,

$$\langle W(\tau) \rangle - \Delta A \approx \beta \sigma^2(W)/2 .$$

$$(5.43)$$

This relation for the dissipated work is exact for a Gaussian work distribution of variance $\sigma^2(W)$.

Broad work distributions have two important consequences: first, the statistics will be poor; and, second, a bias in the estimator of the free energy change, $A(t) - A(0) = -\beta^{-1} \ln \langle \exp(-\beta W(t)) \rangle$, will result in free energy estimates that deviate systematically from the correct free energy difference [10]. Specifically, if the free energy is estimated from N work values Wi drawn at random from the work distribution $p_f(W)$,

$$\Delta A_{\text{est}} = -\beta^{-1} \ln \left(N^{-1} \sum_{i=1}^{N} e^{-\beta W_i} \right)$$

$$(5.44)$$

then the average $\overline{\Delta A_{\text{est}}}$ over repeated sampling of work values does not converge to the correct free energy: $|\overline{\Delta A_{\text{est}}} - \Delta A| > 0$. In essence, broad work distributions result in large statistical and systematic errors. Various estimates of these errors have been discussed [33–35]. As a rough correction, the bias can be estimated by repeatedly drawing the same number N of work values from a Gaussian (or another appropriate distribution) of the same width as the simulation work distribution, and subtracting the numerical estimate of the free energy obtained from (5.44), from the analytical value. That estimated bias can then be added to the estimator (5.44) for the simulation data [15]. Alternatively, the statistics with which rare low W values appear can be used to obtain improved estimators [36, 37].

To reduce the width of the work distributions, the transformation can be staged by breaking it up into segments, as is commonly done in regular free energy perturbation theory. Such an approach has been used, for instance, to calculate the potential of mean force between two solutes in water [20] and similarly in reference [38]. However, dividing up the transformation into multiple segments requires re-equilibration at intermediate values of the coupling parameter.

Cumulant Estimators

Instead of estimating $-\beta^{-1} \ln\langle\exp(-\beta W(t))\rangle$ directly using (5.44), one can use cumulant expansion approaches, as in regular free energy perturbation theory (see e.g., [20, 39] for combining cumulant expansions about the initial and final states). Unbiased estimators for cumulants can be used. Probably the most useful relations involve averages and variances of the work:

$$\Delta A \approx \langle W \rangle - \beta \, \sigma^2(W)/2 \tag{5.45}$$

$$\approx -\langle \underline{W} \rangle + \beta \, \sigma^2(\underline{W})/2 \tag{5.46}$$

Results from forward and backward averages can also be combined [20] into symmetric perturbation formulas:

$$\Delta A \approx (\langle W \rangle - \langle \underline{W} \rangle)/2 \tag{5.47}$$

$$\approx (\langle W \rangle - \langle \underline{W} \rangle)/2 - \frac{\beta}{12} \left(\sigma^2(W) - \sigma^2(\underline{W}) \right) \tag{5.48}$$

Note that in this "optimal" estimator, the variances are not just averaged, but their difference is divided by 12 [20, 39]. In general, these formulas will be useful if the work distributions are nearly Gaussian and, correspondingly, the variances of forward and backward work are nearly identical, $\sigma^2(W) \approx \sigma^2(\underline{W})$. As with other perturbation expressions, the lower-order formula using only the two average work values will be more robust, but may result in comparatively larger systematic errors; higher-order formulas, in contrast, yield larger statistical errors. Nevertheless, higher cumulants have been used successfully in some free energy calculations (see, e.g., [39]).

Histogram Analysis

Potentially more accurate free energies can be obtained by using histogram techniques. In particular, the Crooks relation, (5.35), allows us to use Bennett's method of overlapping histograms [21]. If the work distributions from forward and backward paths overlap, or can be extrapolated into the overlap region, we can use (5.35) to extract free energies. In brief, one collects two histograms $p_f(w = W)$ and $p_b(w = -\underline{W})$ of the work values, one each for forward and backward perturbations, with identical bin locations W_i and widths ΔW. The histograms should be properly normalized such that $\sum_i p_f(W_i)\Delta W = \sum_i p_b(W_i)\Delta W = 1$. The logarithm of the ratio of the histogram values, $\ln[pf(w)/pb(w)] = \beta(w - \Delta A)$ plotted against the work w should then give a straight line of slope β that intercepts the $w = 0$ line at a value of $-\beta\Delta A$. This result also provides us with an important test: if the slope deviates significantly from β, the ensembles of forward and backward paths are not mutually consistent, pointing either to sampling issues or other problems.

Bennett's Optimal "Acceptance Ratio" Estimator

Beyond histogram analysis, one can also use integrated forms of the Crooks relation, (5.35), to estimate free energies. Formally, we can rewrite (5.35) as an average

$$\int f(W; \Delta A)e^{-\beta W}p_f(W)\mathrm{d}W = \int f(W; \Delta A)e^{-\beta\Delta A}p_b(W)\mathrm{d}W \tag{5.49}$$

by multiplying both sides of (5.35) with an arbitrary function f(W; ΔA), bringing the denominator of the left-hand side to the right, and integrating over W. Equation (5.49) is an implicit equation for ΔA, and Bennett [21] determined the function $f = [e^{-\beta(W-\Delta A)}/N_f + 1/N_b]^{-1}$ that minimized the average squared error of the estimated free energy (with N_f and N_b the number of trajectories forward and backward, respectively). An identical result was recently obtained by using a maximum-likelihood approach [22]. The free energy then satisfies the implicit relation

$$\sum_{i=1}^{N_f} \frac{1}{1 + \frac{N_f}{N_b}e^{\beta(W_i - \Delta A)}} = \sum_{i=1}^{N_b} \frac{1}{1 + \frac{N_b}{N_f}e^{\beta(\underline{W}_i + \Delta A)}} \tag{5.50}$$

Note that on the right-hand side, the work $\underline{W}$ is that of the reversed path (i.e., has the opposite sign; see (5.33)). Equation (5.50) has to be solved numerically (e.g., by using a Newton–Raphson solver) for the free-energy difference ΔA.

A word of caution is in place here: the assumption in both Bennett's original calculation [21] and in the maximum-likelihood formalism [22] is that all work values are statistically independent. This may not be the case, for instance, if consecutive initial configurations are close in time along an equilibrium trajectory. If results from either the forward or reverse path are more strongly correlated than those from the opposite path direction, distortions of the free energy estimate seem likely. A typical case where this problem arises is particle insertion versus particle removal [32]. It is generally easier to obtain good statistics for particle insertion than for removal, because large numbers of particles can be inserted into existing configurations but only a given finite number of particles can be removed. Therefore, the insertion results tend to be strongly correlated, but their large number could overwhelm the removal data. In such cases, care is necessary, and (5.50) may not be applicable.

Protocol for Free Energy Estimates from Nonequilibrium Work Averages

Following is a schematic protocol summarizing the main steps in estimating free energies from nonequilibrium work averages.

1. Create equilibrium ensemble of starting configurations. To create N initial conformations representative of the equilibrium ensemble for Hamiltonian $\mathcal{H}$ $(z, \lambda = 0)$, one can, for instance, save conformations at regular intervals during a long equilibrium simulation. In some cases, accelerated sampling procedures may be necessary.

2. Perform nonequilibrium simulations. During the N simulations of duration τ, one for each of the N initial conformations, the Hamiltonian is changed from $\mathcal{H}$ $(z, \lambda(0) = 0)$ to $\mathcal{H}$ $(z, \lambda(\tau) = 1)$.

3. Determine work distribution. After every time (or Monte Carlo) step during the simulations, change the Hamiltonian according to the prescribed path $\lambda(t)$ and accumulate the work performed as part of that change. Collect the work values from the N runs for analysis.

4. Perform backward perturbation to improve accuracy. If possible, repeat the procedure starting from $\mathcal{H}$ $(z, \lambda = 1)$.

5. Analyze work distributions. Use the histogram analysis, optimal and cumulant estimators of Sect. 5.7 to determine free energy difference.

6. Check. Use the Crooks relation, (5.35), to check whether forward and backward work distributions are consistent. Check for consistency of free energies obtained from different estimators. If the amount of dissipated work is large, caution may be necessary. If cumulant expressions are used, the work distributions should be nearly Gaussian, and the variances of forward and backward perturbations should be of comparable size (as required by (5.35) for Gaussian work distributions). Systematic errors from biased estimators should be taken into consideration. Statistical errors can be estimated, for instance, by performing a block analysis.

Illustrating Example

In the following, we will briefly illustrate the application of nonequilibrium free energy calculations for a simple 1D model system. Shown in Fig. 5.1 are the potential energy surfaces

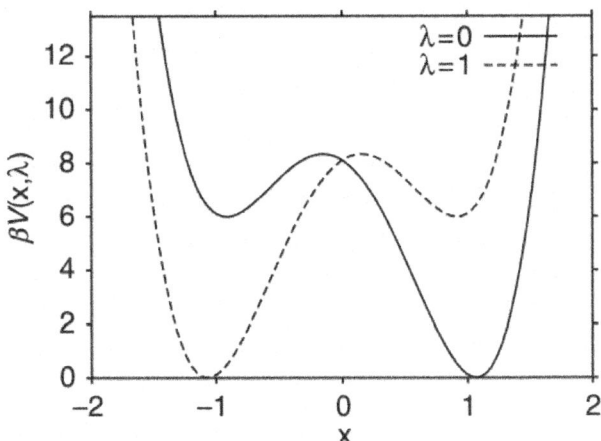

Figure 5.1: Potential energy surface for $\lambda = 0$ (initial state; solid line) and $\lambda = 1$ (final state; dashed line).

$$\beta V(x, \lambda) = 5(x^2 - 1)^2 + 6(\lambda - 1/2)x \tag{5.51}$$

for $\lambda = 0$ and 1. The free energy is defined as $\beta A(\lambda) = -\ln \int \exp[-\beta V(x,\lambda)]dx$. Because of the symmetry with respect to λ [i.e., $V(x,\lambda) = V(-x, 1-\lambda)$], the free-energy difference between the two states $\lambda = 0$ and $\lambda = 1$ vanishes, $\Delta A = A(1) - A(0) = 0$.

We have performed Brownian dynamics simulations (i.e., overdamped Langevin dynamics) on the potential surface $V(x,\lambda)$ with a time-dependent λ. In the simulations, the position x was updated as

$$x(t + \Delta t) = x(t) - D\beta \left.\frac{\partial V(x,t)}{\partial x}\right|_{x=x(t)} \Delta t + (2D\Delta t)^{1/2}g_t \tag{5.52}$$

where g_t are uncorrelated Gaussian random numbers of zero mean and unit variance. The work was accumulated as

$$W(t + \Delta t) = W(t) + V[x(t), t + \Delta t] - V[x(t), t] \tag{5.53}$$

with $W(0) = 0$. The diffusion coefficient was chosen as $D = 1$, with a time step of $\Delta t = 0.001$.

The simulations were started from an equilibrium Boltzmann distribution on the free energy surface for $\lambda = 0$. During a time t = 1, λ was changed linearly in time from 0 to 1. We also performed simulations in the backward direction. However, because of the symmetry of V with respect to λ, backward transformations are equivalent to forward transformations. Along the resulting trajectories, the work βW was accumulated. Figure 5.2 shows the probability distributions of the work on the forward direction, and on the backward direction multiplied by $\exp(-\beta W)$. As expected from (5.35) for $\Delta A = 0$, the two distributions agree nicely.

The work distribution is bi-modal, with a dominant (>99%) peak at high work values of $\beta W \approx 6$, and a small second peak at low work values of $\beta W \approx -6$. The high-work peak is mainly due to trajectories that start in the low-energy well near

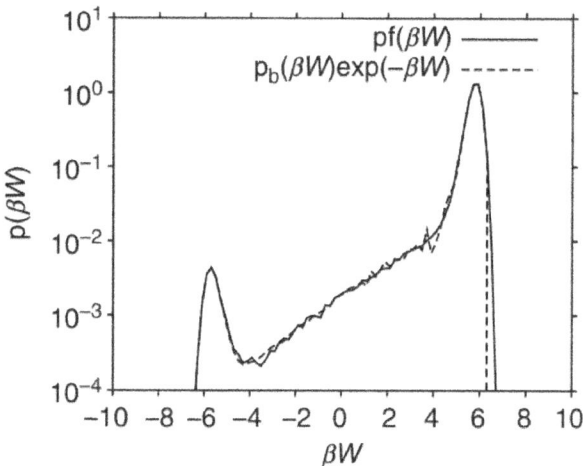

Figure 5.2: Distribution of the work accumulated when transforming the potential surface from $\lambda = 0$ to 1 during a time t = 1. The solid and dashed lines show the distribution of the work on forward and backward paths, respectively. The work distribution for backward paths was multiplied by $\exp(-\beta W)$. The distributions are plotted on a semi-logarithmic scale.

x = 1 (Fig. 5.1). The low-work peak is due to trajectories that start already in the high-energy well at $x \approx -1$.

We used the simulation results to test various estimators for the free energy difference, as shown in Fig. 5.3. Sample sizes N ranged from 100 to 10,000. At least 1,000 independent samples of that size were used to determine the distribution of estimated free energies ΔA_{est}. The resulting histograms of ΔA_{est} are plotted in Fig. 5.3 for different sample sizes and different estimators. Ideally, the histograms should all be delta functions centered at $\Delta A_{est} = 0$. However, for $N = 100$ we find, for instance, a bimodal distribution of free energies estimated by applying the straightforward exponential average, (5.44), and averaging results from forward and backward perturbations (Fig. 5.3a). The asymmetry of the distribution of estimated free energies reflects the strong bias of the exponential estimator, and its width reflects the large statistical error because few events can dominate the estimate. Only for sample sizes of 1,000 to 10,000 does the distribution of ΔA_{est} become centered around zero.

We also used the cumulant estimators of reference [20] in which means (5.47) and variances (5.48) of forward and backward paths are symmetrically combined. As shown in Figs. 5.3b and c, the resulting distributions of estimated free energy differences are sharply centered around $\Delta A_{est} = 0$, thus giving excellent free energy estimates. However, this good agreement is somewhat surprising because the underlying work distributions are highly non-Gaussian, as shown in Fig. 5.2. Indeed, the good performance is due to cancellation of errors because of the symmetry of the problem (such that on average all cumulants exactly cancel each other); for "asymmetric" perturbations from $\lambda = 0$ to 2, the cumulant estimators produce substantial systematic errors (Fig. 5.4).

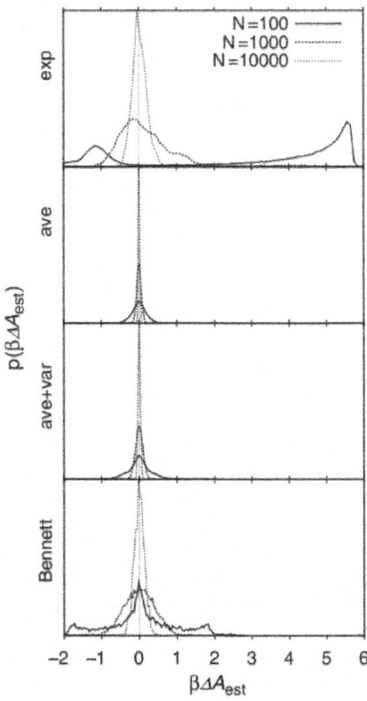

Figure 5.3: Comparison of different free energy estimators. Plotted are distributions of estimated free energies using sample sizes (i.e., number of in-

dependent simulation runs) of $N = 100$ simulations (solid lines), as well as $N = 1,000$ (long dashed) and $N = 10,000$ simulations (short dashed lines). (a) Exponential estimator, (5.44). (b) Cumulant estimator using averages from forward and backward paths, (5.47). (c) Cumulant estimator using averages and variances from forward and backward paths, (5.48). (d) Bennett's optimal estimator, (5.50).

Shown in Fig. 5.3d are free energies estimated from the same forward and backward simulation runs using Bennett's optimal estimator, obtained by solving (5.50) using a Newton–Raphson method. Unlike the direct exponential estimator, we find the optimal estimator to be centered around $\Delta A_{est} = 0$. However, the distribution of the estimated free energies is considerably wider than that of the cumulant estimators, with symmetric wings at the smallest sample size of $N = 100$ (i.e., 50 forward and backward paths each).

In summary, using work collected from forward and backward paths greatly improves the accuracy of the estimates, and for the symmetric system studied here eliminates the bias. In our particular example, the cumulant estimators using forward and backward work data produce the most precise free energy estimates, followed by Bennett's optimal estimator. However, this somewhat poorer performance of the "optimal" estimator is caused in part by the high degree of symmetry of the system studied.

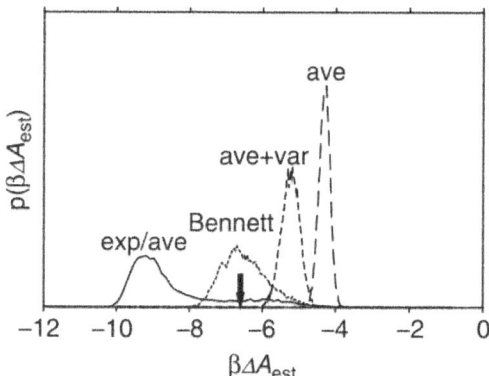

Figure 5.4: Comparison of different free energy estimators for "asymmetric" perturbation from $\lambda = 0$ to 2 within $t = 1$. Shown are distributions of free energies estimated using the direct exponential average, (5.44), averaged over forward and backward perturbations (solid line); averages (5.47) from forward and backward paths (long dashed line); averages and variances (5.48) from forward and backward paths (short dashed line); and Bennett's optimal estimator, (5.50), (dotted line). In all cases, free energies were estimated from $N = 1,000$ simulations. The vertical arrow indicates the actual free energy difference of $\beta\Delta A \approx -6.6$.

Calculating Potentials of Mean Force

One of the objectives in free energy calculations is to obtain potentials of mean force $G(r)$ along a chosen coordinate $r = r(z)$ defined as

$$e^{-\beta G(r)+C} = \int dz e^{-\beta \mathcal{H}_0(z)} \delta(r - r(z))$$

(5.54)

where C is an arbitrary constant. Usually, r is a dynamic, fluctuating variable that depends on the coordinates of the particles in the system. In computer simulations, we can normally use Jarzynski's identity directly by making r a non-fluctuating, externally-controlled coupling parameter, i.e., $\lambda \equiv r$. However, in single-molecule pulling experiments and in their simulation equivalent, steered molecular dynamics [40–43], we may not be able to (or want to) control r directly. In those experiments (or simulations), the system is usually coupled to a moving harmonic spring that drives the coordinate r. An example would be the end-to-end extension of a molecule which is attached to a pulling spring (atomic force microscope, laser optical tweezer, etc.).

When calculating the potential of mean force along a fluctuating coordinate r, we can at best observe r (e.g., the instantaneous molecular extension), but we do not set it explicitly. Therefore, r is no longer an externally controlled coupling parameter, and Jarzynski's identity does not immediately apply. However, as was shown in [3], an extension produces the desired result.

For simplicity, let us consider a molecular system with a Hamiltonian $\mathcal{H}_0(z)$ that is coupled to a harmonic spring with spring constant k and a time-dependent minimum $r(t)$. The explicitly time-dependent Hamiltonian of the complete system is then

$$\mathcal{H}(z, t) = \mathcal{H}_0(z) + \frac{k}{2}(r(z) - r(t))^2$$

(5.55)

where $r(z)$ is the molecular extension and $V(r,t) = k(r(z) - r(t))^2/2$ is the potential energy of the spring.

We can now use the Feynman–Kac relation to obtain the potential of mean force. By multiplying both sides of (5.14) with $\exp[\beta k(r(z) - r(t))^2/2]\delta(r - r(z))$ and integrating over z, we obtain an expression for the potential of mean force

$$e^{-\beta G(r)} = \left\langle \delta[r - r(z(t))]\, e^{-\beta\left(\int_0^t \frac{\partial V\{r[z(\tau)],\tau\}}{\partial \tau} d\tau - V\{r[z(t)],t\}\right)} \right\rangle$$

(5.56)

Formally, by weighting trajectory end points $r(z(t))$ with the Boltzmann factor of the accumulated work (minus the energy stored in the pulling spring), we can produce an unbiased histogram of r proportional to exp($-\beta G(r)$). Practically, r values will cluster near r(t) because of the coupling to the spring, and the histogram will have good statistics only there. However, histograms obtained at different times t can be combined to obtain results over a broad range of r. This approach can be thought of as "dynamic umbrella sampling." By adapting the histogram-reweighting procedure of Ferrenberg and Swendsen [44] for the nonequilibrium averages [3, 15], we obtain the following expression:

$$e^{-\beta G(r)} = \frac{\displaystyle\sum_t \frac{\langle \delta[r - r(t)] \exp(-\beta W(t)) \rangle}{\langle \exp(-\beta W(t)) \rangle}}{\displaystyle\sum_t \frac{\exp[-\beta V(r,t)]}{\langle \exp(-\beta W(t)) \rangle}}$$

(5.57)

Here, the sums are over different time slices t at which coordinates r and corresponding work values have been saved.

Approximate Relations for Potentials of Mean Force

In some cases, one may not be able to collect enough trajectories for a histogram analysis. Instead, one can use moments [15]. In particular, the derivative of the free energy is approximately given by the average force (weighted by the Boltzmann factor of the accumulated work)

$$G_0'(\overline{r(t)}) \approx -\left\langle \frac{\partial V(r,t)}{\partial r} \right\rangle_{r=\overline{r(t)}} .$$

(5.58)

The average coordinate r is defined as:

$$\overline{r(t)} = \frac{\langle r(t)e^{-\beta W(t)} \rangle}{\langle e^{-\beta W(t)} \rangle}$$

(5.59)

This approximation is valid if r(t) is approximately Gaussian distributed.

Above, we obtained the potential of mean force by collecting work and r values. Formally, we can also use the free energy $\Delta A(t)$ measured at all times t to extract the underlying potential of mean force. If we multiply (5.56) with $\exp[-\beta k(r(z) - r(t))^2/2]$ and integrate over r, we obtain

$$e^{-\beta \Delta A(t)} = \int_{-\infty}^{\infty} dr e^{-\beta G(r)} e^{-\beta k(r-r(t))^2/2} .$$

(5.60)

Effectively, this constitutes a Fredholm integral equation of the first kind for $\exp[-\beta G(r)]$ where we "know" the left-hand side, $\exp(-\beta \Delta A(t)) =$

$\langle \exp(-\beta W(t)) \rangle$, from (5.8) and the kernel is given by the Boltzmann factor of the pulling potential. In principle, we could solve the integral equation, for instance, by projection onto orthogonal functions to turn it into a linear equation. However, such approaches are notoriously unstable, and thus not too promising in practice.

For very stiff pulling springs (where r is almost a coupling parameter under external control), we can instead pursue the so-called "stiff-spring approximation" of Park et al. [45]. A Fourier representation of the spring Boltzmann factor on the right-hand side of (5.60) results in

$$e^{-\beta \Delta A(t)} = \int ds \int dr \frac{e^{-\beta G(r) - s^2/(2\beta k) - is(r - r(t))}}{(2\pi \beta k)^{1/2}} .$$
(5.61)

To lowest order in 1/k (with $k \to \infty$), we ignore the s^2 in the exponent and the free-energy offset from the k-dependent numerator, and obtain simply

$$\Delta A(t) \approx G(r = r(t)) .$$
(5.62)

This is the result expected if r is effectively a control parameter. Taylor expansion with respect to 1/k introduces the correction terms as derivatives. To first-order in 1/k, we obtain

$$\Delta A(t) \approx G(r = r(t)) + \frac{1}{2k} \left(\frac{G''(r)}{\beta} - G'(r)^2 \right)_{r=r(t)}$$
(5.63)

To order 1/k, this relation can be inverted, giving the "stiff-spring approximation" derived originally in [45]

$$\Delta G(r = r(t)) \approx A(t) - \frac{1}{2kv^2} \left(\frac{\ddot{A}(t)}{\beta} - \dot{A}(t)^2 \right)$$
(5.64)

where we assumed for simplicity that r(t) = vt with v a constant pulling velocity. (Otherwise, there will be correction terms including $\ddot{z}(t)$.)

One of the major advances in the application of Jarzynski's identity to the calculation of free energies came from coupling it to path sampling [46, 47]. In a typical application with fast switching, the system is rapidly driven out of equilibrium as the coupling parameter is changed, and nearly all trajectories are essentiallyirrelevant for the free energy estimate (i.e., the Boltzmann factor of the trajectory endpoint is small compared to typical values in an equilibrium ensemble). Effectively, those trajectories are wasted. In the biased path sampling of Sun [46], new trajectories are created using a Monte Carlo sampling approach by "perturbing" existing trajectories. To bias the sampling toward important but poorly sampled small work values, an appropriate bias is added to the acceptance criterion. However, despite this appealing feature, a recent efficiency analysis suggests that this biased search in the optimal case is less efficient than an optimally biased Zwanzig-type free energy perturbation [13].

Applications

After some initial studies for model systems [20, 38, 48–51], nonequilibrium work averages have so far been applied most extensively in the context of steered molecular dynamics simulations. In steered molecular dynamics [40–43], the molecular system is coupled to a moving harmonic spring, usually through a group of atoms. This approach is inspired by single-molecule pulling experiments using atomic force microscopes or laser optical tweezers. As the spring is moved, tension builds up in the molecular system, eventually forcing a molecular transition. Examples include the unfolding of proteins [42, 52], dissociation of ligands [40, 41], or dragging molecules through a channel [53, 54]. Usually, one is interested in the potential of mean force along the pulling direction which can be obtained using the dynamic umbrella sampling approach of Sect. 5.9 and [3] or approximations to it.

SUMMARY

Jarzynski's identity [1] provides a new and potentially powerful tool for the determination of free energy differences in computer simulations. As an immediate application, it puts slow-growth free energy calculations [8, 9] on a fully quantitative footing. More importantly, with the help of (5.8), free energy differences can be calculated at arbitrary rates of transformation. Probably the most important applications to date are in the analysis of single-molecule pulling (and related) experiments where nonequilibrium conditions cannot be easily avoided [3–6]. Extensions of Jarzynski-type relations between work distributions and free energy difference to quantum dynamics have been derived [55–57].

Computational efficiency remains a central question. For a given amount of computer time, how "good" are nonequilibrium estimates of free energy differences compared to estimates from equilibrium methods? Overall, evidence is mounting that nonequilibrium methods are less efficient than equilibrium methods [13, 20, 38]. However, new approaches have been suggested that use long time steps [25]. For relatively slow transformations, it has been shown [20] that for a given amount of computer simulation time, one obtains more accurate results for few slow transformationsthan for many fast transformations.

At the other extreme, i.e., in the limit of nearoptimally work-biased fast transformations [13], instantaneous (Zwanzig-type [12]) perturbations appear to be optimal. However, repeated calculations appear to offer some advantages. For instance, if multiple distinct pathways over different "saddles" connect the two states, and equilibration between those paths is slow, then one could imagine that an umbrella sampling approach might get stuck in one "transition path" valley (in particular, if runs in a new umbrella window are initiated from a configuration of the preceding window). By using multiple independent runs, nonequilibrium methods seem more likely to explore all possible pathways, and that could well lead to more accurate free energies when compared to naive applications of standard approaches. However, such vague arguments deserve a more quantitative analysis. Moreover, nonequilibrium methods based on Jarzynski's identity are still a very recent addition to the arsenal for free energy calculations, and one may expect considerable improvements in their implementation as well as new developments [23, 25, 58–65].

The identities of Jarzynski and Crooks, and related expressions, allow us to determine free energy differences rigorously from nonequilibrium simulations. In those simulations, the Hamiltonian is changed continuously (or step wise) from the initial ($\lambda = 0$) to the final energy surface ($\lambda = 1$). Free energies are obtained from the distribution of work accumulated during repeated transformations. The resulting formalism connects free energy perturbation theory with thermodynamic integration (as infinitely fast and slow perturbations, respectively). Moreover, it puts slow-growth methods on a sound theoretical basis. In practical applications, choosing reliable estimators and sampling procedures are critical for the success. The Crooks identity, (5.35), provides a powerful test of sampling and implementation by probing the consistency of work distributions obtained on forward ($0 \rightarrow 1$) and backward ($1 \rightarrow 0$) perturbations. The usefulness of nonequilibrium work methods is most evident in cases where nonequilibrium cannot be avoided easily, for instance in pulling experiments or in simulations where the transformation has to be completed within a given amount of time. In other cases, there is evidence that (near)equilibrium methods produce more accurate results for a given amount of computer time. However, nonequilibrium methods are an area of much ongoing investigation, and we can expect a steady stream of new insights relevant both fundamentally and practically.

REFERENCES

1.   Jarzynski, C., Nonequilibrium Equality for Free Energy Differences, Phys. Rev. Lett. 1997, 78, 2690–2693

2.   Jarzynski, C., Equilibrium Free Energy Differences from Nonequilibrium Measurements. A Master-Equation Approach, Phys. Rev. E 1997, 56, 5018–5035

3.   Hummer, G.; Szabo, A., Free Energy Reconstruction from Nonequilibrium Single-Molecule Pulling Experiments, Proc. Natl. Acad. Sci. USA 2001, 98, 3658–3661

4.   Liphardt, J.; Dumont, S.; Smith, S. B.; Tinoco, I.; Bustamante, C., Equilibrium Information from Nonequilibrium Measurements in an Experimental Test of Jarzynski's Equality, Science 2002, 296, 1832–1835

5.   Noy, A., Direct Determination of the Equilibrium Unbinding Potential Profile for a Short DNA Duplex from Force Spectroscopy Data, Appl. Phys. Lett. 2004, 85, 4792–4794

6.   Trepagnier, E. H.; Jarzynski, C.; Ritort, F.; Crooks, G. E.; Bustamante, C. J.; Liphardt, J., Experimental Test of Hatano and Sasa's Nonequilibrium Steady-State Equality, Proc. Natl. Acad. Sci. USA 2004, 101, 15038–15041

7.   Born, M., Volumen und Hydratationswarme der Ionen, ̈ Z. Phys. 1920, 1, 45–48

8.   Postma, J. P. M.; Berendsen, H. J. C.; Haak, J. R., Thermodynamics of Cavity Formation in Water. A Molecular Dynamics Study, Faraday Symp. Chem. Soc. 1982, 17, 55

9.   Straatsma, T. P.; Berendsen, H. J. C.; Postma, J. P. M., Free Energy of Hydrophobic Hydration. A Molecular Dynamics Study of Noble Gases in Water, J. Chem. Phys. 1986, 85, 6720–6727

10.   Wood, R. H.; Muhlbauer, W. C. F.; Thompson, P. T., Systematic Errors in Free Energy ̈ Perturbation Calculations Due to a Finite Sample of Configuration Space. Sample-Size Hysteresis, J. Phys. Chem. 1991, 95, 6670–6675

11.   Hermans, J., Simple Analysis of Noise and Hysteresis in (Slow-Growth) Free Energy Simulations, J. Phys. Chem. 1991, 95, 9029–9032

12.   Zwanzig, R. W., High-Temperature Equation of State by a Perturbation Method. I. Nonpolar Gases, J. Chem. Phys. 1954, 22, 1420–1426

13.   Oberhofer, H.; Dellago, C.; Geissler, P. L., Biased Sampling of Nonequilibrium Trajectories. Can Fast Switching Simulations Outperform Conventional Free Energy Calculation Methods, J. Phys. Chem. B 2005, 109, 6902–6915

14.   Roepstorff, G., Path Integral Approach to Quantum Physics, Springer: Berlin Heidelberg New York, 1994

15. Hummer, G.; Szabo, A., Free Energy Surfaces from Single-Molecule Force Spectroscopy, Acc. Chem. Res. 2005, 38, 504–513

16. Crooks, G. E., Path-Ensemble Averages in Systems Driven Far from Equilibrium, Phys. Rev. E 2000, 61, 2361–2366

17. Hatano, T.; Sasa, S., Steady-State Thermodynamics of Langevin Systems, Phys. Rev. Lett. 2001, 86, 3463–3466 18. Crooks, G. E., Nonequilibrium Measurements of Free Energy Differences for Microscopically Reversible Markovian Systems, J. Stat. Phys. 1998, 90, 1481–1487

18. Crooks, G. E., Entropy Production Fluctuation Theorem and the Nonequilibrium Work Relation for Free Energy Differences, Phys. Rev. E 1999, 60, 2721–2726

19. Hummer, G., Fast-Growth Thermodynamic Integration Error and Efficiency Analysis, J. Chem. Phys. 2001, 114, 7330–7337

20. Bennett, C. H., Efficient Estimation of Free Energy Differences from Monte Carlo Data, J. Comput. Phys. 1976,

21. 22, 245–268 22. Shirts, M. R.; Bair, E.; Hooker, G.; Pande, V. S., Equilibrium Free Energies from Nonequilibrium Measurements Using Maximum-Likelihood Methods, Phys. Rev. Lett. 2003, 91, 140601

22. Shirts, M. R.; Pande, V. S., Comparison of Efficiency and Bias of Free Energies Computed by Exponential Averaging. The Bennett Acceptance Ratio and Thermodynamic Integration, J. Chem. Phys. 2005, 122, 144107

23. Allen, M. P.; Tildesley, D. J., Computer Simulation of Liquids, Clarendon: Oxford, UK, 1987

24. Lechner, W.; Oberhofer, H.; Dellago, C.; Geissler, P. L., Equilibrium Free Energies from Fast-Switching Trajectories With Large Time Steps, J. Chem. Phys. 2006, 124, 044113

25. Reinhardt, W. P.; Hunter III, J. E., Variational Path Optimization and Upper and Lower Bounds for the Free Energy via Finite Time Minimization of the External Work, J. Chem. Phys. 1992, 97, 1599–1601

26. Miller, M. A.; Reinhardt, W. P., Efficient Free Energy Calculations by Variationally Optimized Metric Scaling Concepts and Applications to the Volume Dependence of Cluster Free Energies and to Solid–Solid Phase Transitions, J. Chem. Phys. 2000, 113, 7035–7046

27. Best, R. B.; Hummer, G., Reaction Coordinates and Rates from Transition Paths, Proc. Natl. Acad. Sci. USA 2005, 102, 6732–6737

28. Chandler, D., Statistical Mechanics of Isomerization Dynamics in Liquids and the Transition State Approximation, J. Chem. Phys. 1978, 68, 2959–2970

29. Hummer, G., From Transition Paths to Transition States and Rate Coefficients, J. Chem. Phys. 2004, 120, 516–523

30. Jarzynski, C., Rare Events and the Convergence of Exponentially Averaged Work Values, Phys. Rev. E 2006, 73, 046105

31. Frenkel, D., Free-Energy Computation and First-Order Phase Transitions. In Molecular Dynamics Simulations of Statistical Mechanical Systems. Proceedings of the Enrico Fermi Summer School, Varenna, 1985 (Amsterdam, 1986), Ciccotti, G.; Hoover, W. G., Eds., North-Holland, pp. 151–188

32. Gore, J.; Ritort, F.; Bustamante, C., Bias and Error in Estimates of Equilibrium Free-Energy Differences from Nonequilibrium Measurements, Proc. Natl. Acad. Sci. USA 2003, 100, 12564–12569

33. Zuckerman, D. M.; Woolf, T. B., Theory of a Systematic Computational Error in Free Energy Differences, Phys. Rev. Lett. 2002, 89, 180602

34. Wu, D.; Kofke, D. A., Asymmetric Bias in Free-Energy Perturbation Measurements Using Two Hamiltonian-Based Models, Phys. Rev. E 2004, 70, 066702

35. Zuckerman, D. M.; Woolf, T. B., Overcoming Finite-Sampling Errors in Fast-Switching Free-Energy Estimates. Extrapolative Analysis of a Molecular System, Chem. Phys. Lett. 2002, 351, 445–453

36. Ytreberg, F. M.; Zuckerman, D. M., Efficient Use of Nonequilibrium Measurement to Estimate Free Energy Differences for Molecular Systems, J. Comp. Chem. 2004, 25, 1749–1759

37. Rodriguez-Gomez, D.; Darve, E.; Pohorille, A., Assessing the Efficiency of Free Energy Calculation Methods, J. Chem. Phys. 2004, 120, 3563–3578

38. Hummer, G.; Szabo, A., Calculation of Free Energy Differences from Computer Simulations of Initial and Final States, J. Chem. Phys. 1996, 105, 2004–2010

39. Grubmuller, H.; Heymann, B.; Tavan, P., Ligand Binding Molecular Mechanics Calcu- ¨ lation of the Streptavidin Biotin Rupture Force, Science 1996, 271, 997–999

40. Izrailev, S.; Stepaniants, S.; Balsera, M.; Oono, Y.; Schulten, K., Molecular Dynamics Study of Unbinding of the Avidin-Biotin Complex, Biophys. J. 1997, 72, 1568–1581

41. Paci, E.; Karplus, M., Forced Unfolding of Fibronectin Type 3 Modules. An Analysis by Biased Molecular Dynamics Simulations, J. Mol. Biol. 1999, 288, 441–459

42. Park, S.; Schulten, K., Calculating Potentials of Mean Force from Steered Molecular Dynamics Simulations, J. Chem. Phys. 2004, 120, 5946–5961

43. Ferrenberg, A. M.; Swendsen, R. H., Optimized Monte Carlo Data Analysis, Phys. Rev. Lett. 1989, 63, 1195–1198

44. Park, S.; Khalili-Araghi, F.; Tajkhorshid, E.; Schulten, K., Free Energy Calculation from Steered Molecular Dynamics Simulations Using Jarzynski's Equality, J. Chem. Phys. 2003, 119, 3559–3566

45. Sun, S. X., Equilibrium Free Energies from Path Sampling of Nonequilibrium Trajectories, J. Chem. Phys. 2003, 118, 5769–5775

46. Ytreberg, F. M.; Zuckerman, D. M., Single-Ensemble Nonequilibrium Path-Sampling Estimates of Free Energy Differences, J. Chem. Phys. 2004, 120, 10876–10879

47. Hendrix, D. A.; Jarzynski, C., A Fast Growth Method of Computing Free Energy Differences, J. Chem. Phys. 2001, 114, 5974–5981

48. Hummer, G., Fast-Growth Thermodynamic Integration Results for Sodium Ion Hydration, Mol. Simul. 2002, 28, 81–90

49. Hu, H.; Yun, R. H.; Hermans, J., Reversibility of Free Energy Simulations Slow Growth May Have a Unique Advantage With a Note on Use of Ewald Summation, Mol. Simul. 2002, 28, 67–80

50. Darve, E.; Pohorille, A., Calculating Free Energies Using Average Force, J. Chem. Phys. 2001, 115, 9169–9183

51. Marszalek, P. E.; Lu, H.; Li, H. B.; Carrion-Vazquez, M.; Oberhauser, A. F.; Schulten, K.; Fernandez, J. M., Mechanical Unfolding Intermediates in Titin Modules, Nature 1999, 402, 100–103

52. Jensen, M. O.; Park, S.; Tajkhorshid, E.; Schulten, K., Energetics of Glycerol Conduction Through Aquaglyceroporin Glpf, Proc. Natl. Acad. Sci. USA 2002, 99, 6731–6736

53. Amaro, R.; Luthey-Schulten, Z., Molecular Dynamics Simulations of Substrate Channeling Through An Alpha-Beta Barrel Protein, Chem. Phys. 2004, 307, 147–155

54. Mukamel, S., Quantum Extension of the Jarzynski Relation Analogy With Stochastic Dephasing, Phys. Rev. Lett. 2003, 90, 170604

55. Jarzynski, C.; Wojcik, D. K., Classical and Quantum Fluctuation Theorems for Heat Exchange, Phys. Rev. Lett. 2004, 92, 230602

56. De Roeck, W.; Maes, C., Quantum Version of Free-Energy-Irreversible-Work Relations, Phys. Rev. E 2004, 69, 026115

57. Atilgan, E.; Sun, S. X., Equilibrium Free Energy Estimates Based on Nonequilibrium Work Relations and Extended Dynamics, J. Chem. Phys. 2004, 121, 10392–10400

58. Ytreberg, F. M.; Zuckerman, D. M., Peptide Conformational Equilibria Computed Via a Single-Stage Shifting Protocol, J. Phys. Chem. B 2005, 109, 9096–9103

59. Chernyak, V.; Chertkov, M.; Jarzynski, C., Dynamical Generalization of Nonequilibrium Work Relation, Phys. Rev. E 2005, 71, 025102

60. Rodinger, T.; Pomes, R., Enhancing the Accuracy the Efficiency and the Scope of Free ` Energy Simulations, Curr. Opin. Struct. Biol. 2005, 15, 164–170

61. De Koning, M., Optimizing the Driving Function for Nonequilibrium Free-Energy Calculations in the Linear Regime. A Variational Approach, J. Chem. Phys. 2005, 122, 104106

62. Lua, R. C.; Grosberg, A. Y., Practical Applicability of the Jarzynski Relation in Statistical Mechanics. A Pedagogical Example, J. Phys. Chem. B 2005, 109, 6805–6811

63. Adib, A. B., Entropy and Density of States from Isoenergetic Nonequilibrium Processes, Phys. Rev. E 2005, 71, 056128

64. Collin, D.; Ritort, F.; Jarzynski, C.; Smith, S.B.; Tinoco, I.; Bustamante, C. Verification of the Crooks Fluctuation Theorem and Recovery of RNA Folding Free Energies. Nature 2005, 437, 231–234

Chapter 6

NON-UNIFORM FFT AND ITS APPLICATIONS IN PARTICLE SIMULATIONS

Yong-Lei Wang[1], Fredrik Hedman[2], Massimiliano Porcu[1,3], Francesca Mocci[1,3], Aatto Laaksonen[1]

[1]Department of Materials and Environmental Chemistry, Arrhenius Laboratory, Stockholm University, Stockholm, Sweden

[2]Noruna AB, Stockholm, Sweden

[3]Dipartimento di Scienze Chimiche e Geologiche, Cagliari University, Cittadella Universitaria di Monserrato (CA), Monserrato, Italy

ABSTRACT

Ewald summation method, based on Non-Uniform FFTs (ENUF) to compute the electrostatic interactions and forces, is implemented in two different particle simulation schemes to model molecular and soft matter, in classical all-atom Molecular Dynamics and in Dissipative Particle Dynamics for coarse-grained particles. The method combines the traditional Ewald method with a non-uniform fast Fourier transform library (NFFT), making it highly efficient. It scales linearly with the number of particles as $\mathcal{O}(N \log N)$, while being both robust and accurate. It conserves both energy and the momentum to float point accuracy. As demonstrated here, it is straight- forward to implement the method in existing computer simulation codes to treat the electrostatic interactions either between point-charges or charge distributions. It should be an attractive alternative to mesh-based Ewald methods.

INTRODUCTION

Computer modeling and simulation of molecular systems is an established discipline rapidly expanding and widely used from Materials Science to Biological systems. At finite temperatures and normal pressures, molecular compounds are in continuous rapid motion. In liquids and solutions, molecules collide and interact with each other at close distances. They also interact with external fields. In computer simulations of molecular systems, mechanistic ball and spring models are commonly employed to give a simple picture of atoms with specific sizes and masses bonded together with covalent bonds. Molecular equilibrium geometries and interactions are defined by so-called force fields where a somewhat arbitrary division between intramolecular and intermolecular interactions is made, because this partitioning is a simplification of the more detailed and fundamental understanding furnished by a quantum mechanical description.

Intramolecular interactions are normally described by bond stretching, angle bending and torsional angle motion terms. These interactions involve the closest bonded atoms described by two-, three-, and four-body terms, respectively. The bond and angle terms are normally given as harmonic wells while the torsion term is most often expressed as a Fourier sum.

Intermolecular interactions are those between separate molecules but also include all interactions within the same molecule beyond the bonded interactions. Non-bonded interactions are further divided between short-ranged and long-ranged interactions. The short-range interactions mimic the Van der Waals type of forces, while the long-range interactions are electrostatic interactions. The electron distributions around atoms are approxi- mated by fixed point charges, and their interactions are treated by using Coulomb's law.

The effective range of short-range interactions is limited to a specified cut-off. By assuming a uniform beyond the cut-off, correction terms to the energy can be obtained by integrating over the non-zero part of the inter- actions [1-3]. Thus the fast-decaying short-range interactions can be accurately approximated by truncation at the cut-off distance in most calculations.

Artificially collecting and dividing the diffuse and fluctuating electron densities inside and around molecules on single atomic sites are a crude but conceptually simple and effective approximation, since Coulomb's law can be invoked. However, this simplification comes at a price since the interactions between point charges stretch over very long distances. Furthermore, these long-range interactions cannot be truncated without introducing simulation artifacts [4-7].

As the system size grows, calculating the electrostatic interactions becomes the major computational bottleneck. Methods based on Ewald summation [8,9] are still considered as the most reliable choice, and a large variety of schemes to compute them in computer simulations have been proposed [10-16]. There is also a multitude of alternative methods for representing electrostatic interactions. Examples are methods based on a cut-off [17-19], tree and multipole based methods [20-25], multigrid methods [26-28], reaction field methods [29-31], the particle mesh method [32,33] and the isotropic sum method [34].

In this paper, we describe an approach to Ewald summation based on the non-uniform fast Fourier transform technique. We use the acronym ENUF-Ewald summation using Non-Uniform fast Fourier transform (NFFT) technique. Our method combines the traditional Ewald summation technique with the NFFT to calculate electrostatic energies and forces in molecular computer simulations. In the paper, we show that ENUF is an easy-to-implement, practical, and efficient method for calculating electrostatic interactions. Energy and momentum are both conserved to float point accuracy. By a suitable choice of parameters, ENUF can be made to behave as traditional Ewald summation but at the same time gives a computational complexity of $\mathcal{O}(N \log N)$, where N is the number of electrostatic interaction sites in the system. Weighing all these properties together, we believe that ENUF should be an attractive alternative in simulations where the high accuracy of Ewald summation is desired.

In the next section, we summarize the basic methodology to apply Ewald summation to computing the electrostatic energies and forces within periodic boundary conditions. In Section 3, we introduce the Ewald method where the reciprocal space part is calculated based on non-uniform FFTs and discuss the underlying concepts and use of the libraries. We also give general guidelines to implement the method in existing simulation programs. In Section 4, we discuss its implementation in a general purpose atomistic Molecular Dynamics simulation package M.DynaMix giving its scaling characteristics in a standard desktop computer. In Section 5, we demonstrate its implementation in Dissipative Particle Dynamics package. This method is applied to simulating soft charged mesoscopic particles. It is necessary to use charge distributions in order to avoid non-physical aggregation of soft charged particles if point charges are used. This issue is discussed further in Section 5.

ELECTROSTATIC INTERACTIONS

We start by describing a model system of charged particles which captures the most salient features of electrostatic interactions in general MD systems. The electrostatic potential of a system with periodic boundary conditions (PBC) is first stated; we follow with the manipulation of the basic formulas to the form in which they are commonly written; this Section ends with a summary of expressions for both energy and forces.

Ewald Summation

Consider a cubic simulation box with edge length L, containing N charged particles, each with a charge q_i, located at r_i. The boundary conditions in a system without cut-off is represented by replicating the simulation box in all directions. The total electrostatic potential energy of the charge-charge interactions is then given by

$$U_{qq} = \sum_{i=1}^{N} \sum_{j=i+1}^{N} \frac{q_i q_j}{|r_{ij}|} + \frac{1}{2} \sum_{n \neq 0} \sum_{i=1}^{N} \sum_{j=1}^{N} \frac{q_i q_j}{|r_{ij} + nL|} = \frac{1}{2} \sum_{n}^{\dagger} \sum_{i=1}^{N} \sum_{j=1}^{N} \frac{q_i q_j}{|r_{ij} + nL|},$$

(1)

where $r_{ij} = r_i - r_j$, and $|r|$ denotes the length (2-norm) of the vector r. Because of the long-range nature of the electrostatic interactions, U_{qq} includes contributions from all replicas, but exclude self-interactions, which is expressed in the triple sum in Equation (1). The outer sum is taken over all integer vectors $n = (n_1, n_2, n_3)$ such that each $n_i \in Z$.

The $\dagger$ symbol on the first summation sign in Equation (1) indicates that the self-interaction terms should not be included, i.e., when $n = 0$ then the $i = j$ terms are omitted.

The sum in Equation (1) is not an absolutely convergent series, but rather conditionally convergent. As a consequence, the order of summation affects the value of the series. In fact, it was discovered by Riemann that any conditionally convergent series of real terms can be rearranged to yield a series which converges to any prescribed sum [35]. In a sense, this is a situation very similar to the case when a linear equation has an infinite number of solutions because it is under-determined; by adding a set of conditions a unique solution may be defined. For the specific case of Equation (1), a physically relevant summation order has to be prescribed and the boundary conditions of the surrounding media have to be specified.

The lattice sum of Equation (1) can be calculated by a method that was first developed in 1921 by Ewald [8]. He used it to calculate lattice potentials in solids. In the context of Molecular Dynamics, there are several different derivations of the Ewald summation method that are more recent; a small selection is given by [3,9]. In the following discussion we mainly follow the work of de Leeuw et al. [9,36,37].

In [9], de Leeuw et al. developed a technique using convergence factors that transforms the sum of a conditionally convergent series into a series with a well-defined sum. Furthermore, they showed that applying a specific convergence factor is equivalent to a certain summation order. Assuming an overall charge neutral system, $\sum q_i = 0$, and summing the terms in Equation (1) over all integer vectors n in concentric spherical order, they showed that the electrostatic potential energy can be written as

$$U_{qq}(\epsilon_r) = \sum_{1 \le i < j \le N} \frac{q_i q_j}{L} \Psi\left(\frac{r_{ij}}{L}\right) + \frac{\xi}{2L}\sum_{i=1}^{N} q_i^2 + \frac{2\pi}{(2\epsilon_r + 1)L^3}\left(\sum_i q_i r_i\right)^2,$$

(2)

when the surrounding media of the periodically replicated cell is a uniform dielectric with dielectric constant ϵ_r and distances are calculated with the minimum image convention. When the surrounding media is a conductor $(\epsilon_r = \infty)$, the energy can be written as

$$U_{qq}(\epsilon_r = \infty) = \sum_{1 \le i < j \le N} \frac{q_i q_j}{L} \Psi\left(\frac{r_{ij}}{L}\right) + \frac{\xi}{2L}\sum_{i=1}^{N} q_i^2 = U_{qq}(\epsilon_r) - \frac{2\pi}{(2\epsilon_r + 1)L^3}\left(\sum_i q_i r_i\right)^2.$$

(3)

From Equation (3) it is clear that the boundary conditions, vacuum or conductor, have an effect on the energy of the system. Depending on the simulated system and the properties of interest, the choice of boundary conditions can affect the obtained results [38,39].

The function Ψ is given by

$$\Psi(r) = \sum_n \frac{\text{erfc}(\alpha|r+n|)}{|r+n|} + \frac{1}{\pi}\sum_{n \ne 0} \frac{\exp\left(2\pi i n \cdot r - \pi^2 |n|^2 / \alpha^2\right)}{|n|^2},$$

(4)

with the two error functions defined as $\text{erfc}(x) = 1 - \frac{2}{\sqrt{\pi}}\int_0^x e^{-t^2} dt = \frac{2}{\sqrt{\pi}}\int_x^\infty e^{-t^2} dt$ and $\text{erf}(x) = 1 - \text{erfc}(x)$. The number ξ used in Equations (2) and (3) is defined as

$$\xi = \sum_{n \ne 0} \frac{\text{erfc}(\alpha|n|)}{|n|} + \frac{1}{\pi}\sum_{n \ne 0} \frac{e^{-\pi^2 |n|^2 / \alpha^2}}{|n|^2} - \frac{2\alpha}{\sqrt{\pi}},$$

(5)

where $\alpha > 0$ is a free parameter.

Equations (2) and (3) are not in a form that is appropriate for efficient numerical calculations and in the case of Molecular Dynamics simulation we also need expressions for the forces. To arrive at a more suitable form we make the necessary analysis for the electrostatic energy and forces in the following Sections.

Energies in Ewald Summation

To rearrange and expand Equation (2) we first insert r_{ij}/L in Equation (4)

$$\Psi\left(\frac{r_{ij}}{L}\right) = \sum_n \frac{\mathrm{erfc}\left(\alpha\left|\frac{r_{ij}}{L}+n\right|\right)}{\left|\frac{r_{ij}}{L}+n\right|} + \frac{1}{\pi}\sum_{n\neq 0}\frac{\exp\left(2\pi\imath n\cdot\frac{r_{ij}}{L}-\pi^2|n|^2/\alpha^2\right)}{|n|^2}$$

$$= L\sum_n \frac{\mathrm{erfc}\left(\frac{\alpha}{L}\left|r_{ij}+nL\right|\right)}{\left|r_{ij}+nL\right|} + \frac{1}{\pi}\sum_{n\neq 0}\frac{\exp\left(\frac{2\pi\imath}{L}n\cdot r_{ij}-\pi^2|n|^2/\alpha^2\right)}{|n|^2}.$$

(6)

Next we rescale α by making the substitution $\alpha\rightarrow\alpha L$ in Equations (5) and (6) to get

$$\Psi\left(\frac{r_{ij}}{L}\right) = L\sum_n \frac{\mathrm{erfc}\left(\alpha\left|r_{ij}+nL\right|\right)}{\left|r_{ij}+nL\right|} + \frac{1}{\pi}\sum_{n\neq 0}\frac{\exp\left(\frac{2\pi\imath}{L}n\cdot r_{ij}-\pi^2|n|^2/(\alpha L)^2\right)}{|n|^2}$$

(7)

and

$$\xi = \sum_{n\neq 0}\frac{\mathrm{erfc}\left(\alpha|n|L\right)}{|n|} + \frac{1}{\pi}\sum_{n\neq 0}\frac{e^{-\pi^2|n|^2/(\alpha L)^2}}{|n|^2} - \frac{2\alpha L}{\sqrt{\pi}}.$$

(8)

Inserting Equations (7) and (8) into Equation (2) we get

$$U_{qq}(\epsilon_r) = \sum_{1\leq i<j\leq N} q_i q_j \left\{\sum_n \frac{\mathrm{erfc}\left(\alpha\left|r_{ij}+nL\right|\right)}{\left|r_{ij}+nL\right|} + \frac{1}{\pi L}\sum_{n\neq 0}\frac{\exp\left(\frac{2\pi\imath}{L}n\cdot r_{ij}-\pi^2|n|^2/(\alpha L)^2\right)}{|n|^2}\right\}$$

$$+ \frac{1}{2L}\sum_i q_i^2 \left[\sum_{n\neq 0}\frac{\mathrm{erfc}\left(\alpha|n|L\right)}{|n|} + \frac{1}{\pi}\sum_{n\neq 0}\frac{e^{-\pi^2|n|^2/(\alpha L)^2}}{|n|^2} - \frac{2\alpha L}{\sqrt{\pi}}\right] + \frac{2\pi}{(2\epsilon_r+1)L^3}\left(\sum_i q_i r_i\right)^2.$$

(9)

Note that the summation in Equation (9) is for $i<j$ in the first sum. We make further simplifications by studying the terms on the right hand side of Equation (9) for $n=0$ and $n\neq 0$.

When $n=0$ we have the following terms

$$\sum_{1\leq i<j\leq N} q_i q_j \frac{\mathrm{erfc}\left(\alpha|r_{ij}|\right)}{|r_{ij}|} - \frac{\alpha}{\sqrt{\pi}}\sum_i q_i^2 + \frac{2\pi}{(2\epsilon_r+1)L^3}\left(\sum_i q_i r_i\right)^2$$

$$= \frac{1}{2}\sum_{i,j}^{\dagger}\frac{q_i q_j}{|r_{ij}|}\mathrm{erfc}\left(\alpha|r_{ij}|\right) - \frac{\alpha}{\sqrt{\pi}}\sum_i q_i^2 + \frac{2\pi}{(2\epsilon_r+1)L^3}\left(\sum_i q_i r_i\right)^2$$

(10)

with the terms independent of n included. The $\dagger$ symbol indicates that the $i = j$ terms are excluded from the daggered sum.

When ■■ we get

$$
\begin{aligned}
\sum_{1 \leq i < j \leq N} q_i q_j &\left\{ \sum_{n \neq 0} \frac{\mathrm{erfc}\left(\alpha \left| r_{ij} + nL \right|\right)}{\left| r_{ij} + nL \right|} + \frac{1}{\pi L} \sum_{n \neq 0} \frac{\exp\left(\frac{2\pi \iota}{L} n \cdot r_{ij} - \pi^2 |n|^2 / (\alpha L)^2 \right)}{|n|^2} \right\} \\
&+ \frac{1}{2L} \sum_i q_i^2 \left\{ \sum_{n \neq 0} \frac{\mathrm{erfc}\left(\alpha |n| L\right)}{|n|} + \frac{1}{\pi} \sum_{n \neq 0} \frac{e^{-\pi^2 |n|^2 / (\alpha L)^2}}{|n|^2} \right\} \\
&= \frac{1}{2} \sum_{i,j} q_i q_j \sum_{n \neq 0} \left\{ \frac{\mathrm{erfc}\left(\alpha \left| r_{ij} + nL \right|\right)}{\left| r_{ij} + nL \right|} + \frac{1}{\pi L} \frac{e^{-\pi^2 |n|^2 / (\alpha L)^2}}{|n|^2} \exp\left(\frac{2\pi \iota}{L} n \cdot r_{ij} \right) \right\}.
\end{aligned}
$$

$$(11)$$

The factor 1/2 in Equation (11) comes from changing the summation $i < j$ to all pairs i and j, and using the symmetry induced by $r_{ij} = -r_{ji}$ and $\pm n$.

By combining Equations (10) and (11) we identify the real-space term U_{qq}^{real}, the reciprocal-space term U_{qq}^{recip}, and the boundary-condition term U_{qq}^{bc}. The real-space term is given by

$$
\begin{aligned}
U_{qq}^{real} &= \sum_{1 \leq i < j \leq N} q_i q_j \frac{\mathrm{erfc}\left(\alpha |r_{ij}|\right)}{|r_{ij}|} + \sum_{i<j} \sum_{n \neq 0} \frac{\mathrm{erfc}\left(\alpha \left| r_{ij} + nL \right|\right)}{\left| r_{ij} + nL \right|} + \frac{1}{2} \sum_i q_i^2 \sum_{n \neq 0} \frac{\mathrm{erfc}\left(\alpha |n| L\right)}{|n| L} \\
&= \frac{1}{2} \sum_n^{\dagger} \sum_{i,j} \frac{q_i q_j}{r_{ij} + nL} \mathrm{erfc}\left(\alpha \left| r_{ij} + nL \right|\right),
\end{aligned}
$$

$$(12)$$

and the reciprocal-space term is

$$
U_{qq}^{recip} = \frac{1}{\pi L} \sum_{i<j} \sum_{n \neq 0} q_i q_j \frac{e^{-\pi^2 |n|^2 / (\alpha L)^2}}{|n|^2} \exp\left(\frac{2\pi \iota}{L} n \cdot r_{ij} \right) + \frac{1}{2\pi L} \sum_i \sum_{n \neq 0} q_i q_i \frac{e^{-\pi^2 |n|^2 / (\alpha L)^2}}{|n|^2}
$$

$$(13)$$

However, with the symmetries generated by $r_{ij} = -r_{ji}$ and $\pm n$, we get

$$
\begin{aligned}
U_{qq}^{recip} &= \frac{1}{2\pi L} \sum_{i \neq j} \sum_{n \neq 0} q_i q_j \frac{e^{-\pi^2 |n|^2 / (\alpha L)^2}}{|n|^2} \exp\left(\frac{2\pi \iota}{L} n \cdot r_{ij} \right) + \frac{1}{2\pi L} \sum_i \sum_{n \neq 0} q_i q_i \frac{e^{-\pi^2 |n|^2 / (\alpha L)^2}}{|n|^2} \\
&= \frac{1}{2\pi L} \sum_{n \neq 0} \sum_{i,j} q_i q_j \frac{e^{-\pi^2 |n|^2 / (\alpha L)^2}}{|n|^2} \exp\left(\frac{2\pi \iota}{L} n \cdot r_{ij} \right).
\end{aligned}
$$

$$(14)$$

Furthermore, we have

$$U_{qq}^{self} = \frac{\alpha}{\sqrt{\pi}} \sum_i q_i^2 ,$$

(15)

$$U_{qq}^{bc} = \frac{2\pi}{(2\epsilon_r + 1)L^3} \left(\sum_i q_i r_i \right)^2 ,$$

(16)

and finally

$$U_{qq}(\epsilon_r) = U_{qq}^{real} + U_{qq}^{recip} - U_{qq}^{self} + U_{qq}^{bc}.$$

(17)

The reciprocal-space part, Equation (14), can be expanded in two different forms. The first form is in terms of the structure factor,

$$S(n) = \sum_i q_i \exp\left(-\frac{2\pi\iota}{L} n \cdot r_i \right),$$

(18)

and is given by

$$\begin{aligned} U_{qq}^{recip} &= \frac{1}{2\pi L} \sum_{n\neq 0} \frac{e^{-\pi^2 |n|^2/(\alpha L)^2}}{|n|^2} \sum_{i,j} q_i q_j \exp\left(\frac{2\pi\iota}{L} n \cdot (r_i - r_j) \right) \\ &= \frac{1}{2\pi L} \sum_{n\neq 0} \frac{e^{-\pi^2 |n|^2/(\alpha L)^2}}{|n|^2} \sum_i q_i \exp\left(\frac{2\pi\iota}{L} n \cdot r_i \right) \sum_j q_j \exp\left(\frac{2\pi\iota}{L} n \cdot (-r_j) \right) \\ &= \frac{1}{2\pi L} \sum_{n\neq 0} \frac{e^{-\pi^2 |n|^2/(\alpha L)^2}}{|n|^2} S(-n) S(n). \end{aligned}$$

(19)

Now for a fixed, the structure factor n is just a complex number S(n), and the simple fact that

$$a\bar{a} = \mathrm{Re}(a)^2 + \mathrm{Im}(a)^2,$$ gives the real form of Equation (14)

$$\begin{aligned} U_{qq}^{recip} &= \frac{1}{2\pi L} \sum_{n\neq 0} \frac{e^{-\pi^2 |n|^2/(\alpha L)^2}}{|n|^2} \sum_{i,j} q_i q_j \exp\left(\frac{2\pi\iota}{L} n \cdot (r_i - r_j) \right) \\ &= \frac{1}{2\pi L} \sum_{n\neq 0} \frac{e^{-\pi^2 |n|^2/(\alpha L)^2}}{|n|^2} \left[\left(\sum_i q_i \cos\left(\frac{2\pi}{L} n \cdot r_i \right) \right)^2 + \left(\sum_i q_i \sin\left(\frac{2\pi}{L} n \cdot r_i \right) \right)^2 \right]. \end{aligned}$$

(20)

The first form in Equation (19) is used in our fast approach to calculating the reciprocal-space part. The last form in Equation (20) is the most common point-of-departure when implementing the reciprocal-space part.

Forces in Ewald Summation

Now that we have calculated the electrostatic energy of the system we can easily compute the electrostatic forces F_i that act on each particle i. Splitting the forces in the same way as we have split the energy and using Equation (17) we get the total electrostatic force by finding the negative of the gradient of the electrostatic energy

$$F_i = -\nabla_i U_{qq} = -\nabla_i \left[U_{qq}^{real} + U_{qq}^{recip} - U_{qq}^{self} + U_{qq}^{bc} \right] = F_i^{real} + F_i^{recip} - 0 + F_i^{bc}. \tag{21}$$

The subscript i on the ∇ operator indicates that we take the partial derivatives with respect to the position of particle r_i and the 0 in Equation (21) comes from the self-interaction term in Equation (15) being independent of r_i. Before we do this calculation we note a couple basic, but helpful, formulas for calculating derivatives

$$\nabla_i \mathbf{n} \cdot \mathbf{r}_{ij} = \nabla_i \mathbf{n} \cdot (\mathbf{r}_i - \mathbf{r}_j) = \mathbf{n}$$

$$\nabla_i \frac{1}{|\mathbf{r}_{ij}|} = -\frac{\mathbf{r}_{ij}}{|\mathbf{r}_{ij}|^3}$$

$$\nabla_i |\mathbf{r}_{ij}| = \frac{\mathbf{r}_{ij}}{|\mathbf{r}_{ij}|}$$

$$\nabla_i \operatorname{erfc}(\alpha|\mathbf{r}_{ij}|) = -\frac{2\alpha}{\sqrt{\pi}} \frac{\mathbf{r}_{ij}}{|\mathbf{r}_{ij}|} \exp\left(-\alpha^2 |\mathbf{r}_{ij}|^2\right)$$

$$\nabla_i \exp\left(\frac{2\pi\iota}{L} \mathbf{n} \cdot \mathbf{r}_{ji}\right) = -\mathbf{n}\frac{2\pi\iota}{L} \exp\left(\frac{2\pi\iota}{L} \mathbf{n} \cdot \mathbf{r}_{ji}\right)$$

$$\nabla_i \exp\left(\frac{2\pi\iota}{L} \mathbf{n} \cdot \mathbf{r}_{ij}\right) = +\mathbf{n}\frac{2\pi\iota}{L} \exp\left(\frac{2\pi\iota}{L} \mathbf{n} \cdot \mathbf{r}_{ij}\right) \tag{22}$$

With above formulas, it is straightforward to find the different terms of F_i. The contribution from the real-space term, F_i^{real}, becomes

$$F_i^{real} = -\nabla_i U_{qq}^{real} = -\nabla_i \left[\frac{1}{2} \sum_n \sum_{i,j}^{\dagger} \frac{q_i q_j}{|\mathbf{r}_{ij} + \mathbf{n}L|} \operatorname{erfc}(\alpha|\mathbf{r}_{ij} + \mathbf{n}L|) \right]$$

$$= \sum_n \sum_j^{\dagger} q_i q_j \frac{\mathbf{r}_{ij} + \mathbf{n}L}{|\mathbf{r}_{ij} + \mathbf{n}L|^2} \left[\frac{\operatorname{erfc}(\alpha|\mathbf{r}_{ij} + \mathbf{n}L|)}{|\mathbf{r}_{ij} + \mathbf{n}L|} + \frac{2\alpha}{\sqrt{\pi}} \exp\left(-\alpha^2 |\mathbf{r}_{ij} + \mathbf{n}L|^2\right) \right], \tag{23}$$

such that when $n \neq 0$ include all j and otherwise only $j \neq i$ – "the daggered saviour". Equation (20) is convenient to use when calculating the reciprocal-space contribution because it is expressed in terms of charge locations r_i rather than relative distances r_{ij}. Thus the reciprocal-space force is given by

$$F_i^{recip} = -\nabla_i U_{qq}^{recip}$$

$$= -\nabla_i \left[\frac{1}{2\pi L} \sum_{n \neq 0} \frac{e^{-\pi^2 |n|^2 / (\alpha L)^2}}{|n|^2} \left[\left(\sum_i q_i \cos\left(\frac{2\pi}{L} n \cdot r_i\right) \right)^2 + \left(\sum_i q_i \sin\left(\frac{2\pi}{L} n \cdot r_i\right) \right)^2 \right] \right]$$

$$= \frac{2 q_i}{L^2} \sum_{n \neq 0} n \frac{e^{-\pi^2 |n|^2 / (\alpha L)^2}}{|n|^2} \left[\sin\left(\frac{2\pi}{L} n \cdot r_i\right) Re(S(n)) + \cos\left(\frac{2\pi}{L} n \cdot r_i\right) Im(S(n)) \right].$$

$$(24)$$

Finally, the contribution that depends on the boundary condition

$$F_i^{bc} = -\nabla_i \left[\frac{2\pi}{(2\epsilon_r + 1) L^3} \left(\sum_i q_i r_i \right)^2 \right] = -\frac{4\pi}{(2\epsilon_r + 1) L^3} q_i \sum_j q_j r_j.$$

$$(25)$$

Formulas for Energy and Forces in Ewald Summation

Consider a periodically replicated system, with the central box consisting of N point charges q_i and $\sum q_i = 0$. Assume that the surrounding media at the boundary of the periodically replicated system is a uniform dielectric with dielectric constant ϵ_r. The cubic box has edge length L; each charge q_i is located at r_i, and distances are calculated with the minimum image convention. After expansion and rearrangements of Equation (2), rescaling $\alpha \to \alpha L$ and using symmetries induced by $r_{ij} = -r_{ji}$ and $\pm n$, the total electrostatic energy of the system can be written as

$$U_{qq}(\epsilon_r) = U_{qq}^{real} + U_{qq}^{recip} - U_{qq}^{self} + U_{qq}^{bc}$$

$$(26)$$

with the different terms given by

$$U_{qq}^{real} = \frac{1}{2} \sum_n^{\dagger} \sum_{i,j} \frac{q_i q_j}{|r_{ij} + nL|} \operatorname{erfc}\left(\alpha |r_{ij} + nL|\right),$$

$$(27)$$

$$U_{qq}^{recip} = \frac{1}{2\pi L} \sum_{n \neq 0} \sum_{i,j} q_i q_j \frac{e^{-\pi^2 |n|^2/(\alpha L)^2}}{|n|^2} \exp\left(\frac{2\pi \iota}{L} n \cdot r_{ij}\right)$$

$$= \frac{1}{2\pi L} \sum_{n \neq 0} \frac{e^{-\pi^2 |n|^2/(\alpha L)^2}}{|n|^2} \left[\left(\sum_i q_i \cos\left(\frac{2\pi}{L} n \cdot r_i\right)\right)^2 + \left(\sum_i q_i \sin\left(\frac{2\pi}{L} n \cdot r_i\right)\right)^2\right]$$

$$= \frac{1}{2\pi L} \sum_{n \neq 0} \frac{e^{-\pi^2 |n|^2/(\alpha L)^2}}{|n|^2} S(n) S(-n),$$

$$(28)$$

$$U_{qq}^{self} = \frac{\alpha}{\sqrt{\pi}} \sum_i q_i^2,$$

$$(29)$$

$$U_{qq}^{bc} = \frac{2\pi}{(2\epsilon_r + 1) L^3} \left(\sum_i q_i r_i\right)^2.$$

$$(30)$$

Note that $\alpha > 0$ is a free parameter. The structure factor $S(n)$ is defined as

$$S(n) = \sum_i q_i \exp\left(-\frac{2\pi \iota}{L} n \cdot r_i\right).$$

$$(31)$$

The total electrostatic force, F_i, on each particle is

$$F_i = F_i^{real} + F_i^{recip} + F_i^{bc}.$$

$$(32)$$

Each of the force terms given by

$$F_i^{real} = \sum_n^{\dagger} \sum_j q_i q_j \frac{r_{ij} + nL}{|r_{ij} + nL|^2} \left[\frac{\mathrm{erfc}\left(\alpha |r_{ij} + nL|\right)}{|r_{ij} + nL|} + \frac{2\alpha}{\sqrt{\pi}} \exp\left(-\alpha^2 |r_{ij} + nL|^2\right)\right],$$

$$(33)$$

$$F_i^{recip} = \frac{2q_i}{L^2} \sum_{n \neq 0} n \frac{e^{-\pi^2 |n|^2/(\alpha L)^2}}{|n|^2} \left[\sin\left(\frac{2\pi}{L} n \cdot r_i\right) \sum_j q_j \cos\left(\frac{2\pi}{L} n \cdot r_j\right) - \cos\left(\frac{2\pi}{L} n \cdot r_i\right) \sum_j q_j \sin\left(\frac{2\pi}{L} n \cdot r_j\right)\right],$$

$$(34)$$

$$F_i^{bf} = -\frac{4\pi}{(2\epsilon_r + 1) L^3} q_i \sum_j q_j r_j.$$

$$(35)$$

The positive number α, the so called Ewald convergence parameter, is chosen for computational con- venience. Observe that $\mathrm{erfc}(x) \approx e^{-x^2}$ for large values of x. By choosing α large enough in Equation (27), we can ensure that the only terms that contribute in the real-space sum is when $n=0$. This may be expressed so that all terms with $|n| < n_{cut}$ should be included.

Choose a cut-off in both real-space and reciprocal-space so that the neglected terms in the real-space and reciprocal-space parts are of the same order δ, or less. The truncation in real-space implies that a sufficient number of terms must be included in the reciprocal-space sums, Equation (28).

Given a required accuracy δ, n_{cut} is fixed by

$$e^{-\pi^2 n_{cut}^2 /(\alpha L)^2} \le \delta \Rightarrow n_{cut} \ge \frac{\alpha L}{\pi}\sqrt{-\log(\delta)},$$

(36)

and r_{cut} is determined by

$$\mathrm{erfc}\left(\alpha r_{cut}\right) \approx e^{-\alpha^2 r_{cut}^2} \le \delta \Rightarrow r_{cut} = \frac{\pi n_{cut}}{\alpha^2 L}.$$

(37)

We have two conditions and four parameters. With a required δ we may just as well pick a suitable value for n_{cut} and let the above two equations determine α and r_{cut}.

With an optimal choice of parameters the computational effort of the Ewald method becomes $\mathcal{O}(N^{3/2})$ [36] giving a considerable improvement over the $\mathcal{O}(N^2)$ computational complexity implied by the "infinite" reach of the Coulomb interactions.

ENUF: A FAST METHOD FOR CALCULATING ELECTRO-STATIC INTERACTIONS

In Section 2 we summarized known results and prepared the ground for the development of a fast method for Ewald summation using the discrete nonuniform fast Fourier transform (NDFT).

Discrete Fourier Transforms for Non-Equispaced Data

The fast Fourier transform for nonuniform data-points (NFFT) [40] is a generalization of the FFT [41]. Several similar approaches have been proposed; some examples are [42-50] with comparisons in [47,51,52].

The basic idea of NFFT is to combine the standard FFT and linear combinations of a window function that is well localized in both the spatial domain and the frequency domain. A controlled approximation using a cut-off in the frequency domain and a limited number of terms in the spatial domain results in an aliasing error and a truncation error, respectively. The aliasing errors is controlled by the oversampling factor σ_s, and the truncation error is controlled by the number of terms, m, in the spatial/time approximation. For a number of window functions (Gaussian, B-spline, Sinc-power, Kaiser-Bessel), it has been shown that for a fixed over- sampling factor, $\sigma_s > 1$, the error decays exponentially with m [53].

Problem Definition

We wish to calculate the discrete Fourier transform for nonequispaced data. The problem can be stated as follows. For a finite number of given Fourier coefficients $\hat{f}_k \in C$ with $k \in I_M$ we want to evaluate the trigonometric polynomial $f(x) = \sum_{k \in I_M} \hat{f}_k \exp(-2\pi \imath k x)$ at each of the given nonequispaced points $\left(x_j \in \mathcal{D}^d, j = 0, \cdots, N-1 \right)$.

In the literature, points are often called knots. We use the two terms synonymously.

Obviously, the details of an NDFT depend on the definitions of a sampling set for knots, $\mathcal{D}^d$, and an index space I_M. More in-depth discussions and further details can be found in [53,54]. The presentation that follows is mainly drawn from these sources.

Underlying Concepts

Consider a d-dimensional domain $\mathcal{D}^d$ in which the set of nonequispaced knots, or data points, are located. Let

$$\mathcal{D}^d := \left\{ x = (x_t)_{t=0,\cdots,d-1} \in \mathcal{R}^d : -\frac{1}{2} \leq x_t < \frac{1}{2}, t = 0, \cdots, d-1 \right\},$$

(38)

and the set of N data points $\left(\mathcal{X} := \{ x_j \in \mathcal{D}^d : j = 0, \cdots, N-1 \} \right)$. For the application we have in mind d is usually 2 or 3. Let $\mathcal{F}$ be a function space of trigonometric polynomials with degree $M_t (t = 0, \cdots, d-1)$ in dimension t; the function space $\mathcal{F}$ can be defined as

$$\mathcal{F} := \left\{ f : \mathcal{D}^d \mapsto C \text{ such that } f \in \text{span} \left(e^{-2\pi \imath k \cdot} : k \in I_M \right) \right\}.$$

(39)

The dimension of this function space is $\dim(\mathcal{F}) = M_\Pi$, where $M_\Pi = \prod_{t=0}^{d-1} M_t$. The frequencies $k \in I_M$ with the index set I_M are such that

$$I_M := \left\{ k = (k_t)_{t=0,\cdots,d-1} \in \mathcal{Z}^d : -\frac{M_t}{2} \leq k_t < \frac{M_t}{2}, t = 0,\cdots,d-1 \right\}.$$

(40)

Matrix-Vector Formulation

With these preliminary definitions we carry on with the problem of calculating the discrete Fourier transform for nonequispaced data. For a finite number of given Fourier coefficients $\hat{f}_k \in \mathcal{C}$ with $k \in I_M$ we want to evaluate the trigonometric polynomial $f(x) = \sum_{k \in I_M} \hat{f}_k \exp(-2\pi \imath kx)$ at each of the given nonequispaced knots in $\mathcal{X}$, where the product kx is the usual scalar product of the two vectors k and x as $kx := k_0 x_0 + \cdots + k_{d-1} x_{d-1}$. Consequently, for each $x_j \in \mathcal{X}$, we evaluate $f_j = f(x_j) := \sum_{k \in I_M} \hat{f}_k \exp(-2\pi \imath kx_j)$. This may be reformulated in matrix-vector notation by setting

$$f := \left(f_j \right)_{j=0,\cdots,N-1},$$

$$A := \left(e^{-2\pi \imath kx_j} \right)_{j=0,\cdots,N-1; k \in I_M},$$

$$\hat{f} := \left(\hat{f}_k \right)_{k \in I_M}$$

(41)

and writing $f = A\hat{f}$.

Related Matrix-Vector Products

A number of related NDFT matrix-vector products can also be defined. To write them down we let $\bar{A}$ be the complex conjugate of the elements of the matrix A and $A^H = \bar{A}^T$ the transposed complex conjugate of the matrix A. Using these conventions we can name and summarize the related NDFT matrix-vector products and their component representation as

Regular : $f = A\hat{f}, f_j = \sum_{k \in I_M} \hat{f}_k \exp\left(-2\pi \imath kx_j\right).$

(42)

Adjoint : $\hat{f} = A^H f, \hat{f}_k = \sum_{j=0}^{N-1} \hat{f}_j \exp\left(2\pi \imath kx_j\right).$

(43)

Conjugated $: f = \overline{A}\hat{f}, f_j = \sum_{k \in I_M} \hat{f}_k \exp\left(2\pi\imath kx_j\right).$

$$(44)$$

Transposed $: \hat{f} = A^{\mathrm{T}} f, \hat{f}_k = \sum_{j=0}^{N-1} \hat{f}_j \exp\left(-2\pi\imath kx_j\right).$

$$(45)$$

NDFT, FFT and NFFT

From the different NDFT products written in matrix-vector form, as in Equations (42)-(45), it is clear that it takes $\mathcal{O}(NM_\Pi)$ arithmetic operations to transform between the Fourier-samples and the Fourier-coefficients.

This is simply because the matrix A is $N \times M_\Pi$, with $M_\Pi = \dim(\mathcal{F}) = \prod_{t=0}^{d-1} M_t$.

However, for the special case of $M_t = M_{(t=0,\cdots,d-1)}$ and $N = M^d$ equispaced knots $x_k \, (k \in I_M)$, the Fourier-samples f_k can be calculated from the Fourier-coefficients $\hat{f}_k$ by the fast Fourier transform (FFT) with $\mathcal{O}(N \log N)$ arithmetic operations.

The fast Fourier transform for nonequispaced knots (NFFT) is a generalization of the FFT. The essential idea is that of combining a window function with the standard FFT. The window function is a well localized function in both the space domain and frequency domain. Several different window functions and similar approaches have been proposed. The resulting algorithms are approximate and some of them have been shown to have a computational complexity of $\mathcal{O}(M_\Pi \log M_\Pi + \log(1/\epsilon)N)$, where ϵ is the desired accuracy [53].

Fast Ewald Summation

Using optimal parameters in the Ewald summation method implies that the time to calculate the real-space part and the reciprocal-space part are approximately equal. As the number of particles in the system grows we would like to combine the calculation of the short-range part of the potential with the real-space part. This implies that we need to choose a real-space cut-off about the same size as the short-range cut-off. With this nonoptimal choice, the reciprocal-space parts of the Ewald summation method become the most time-consuming to calculate [55].

To show how a fast Ewald summation approach may be obtained from the regular Ewald method, described in Section 2, we focus on the reciprocal-space parts. In Section 3.1, we gave the details of the discrete Fourier transform (DFT) for data that is nonuniformly spaced (NDFT). Based on these definitions we get a number of useful algorithmic primitives. First we reformulate the reciprocal-space part of the regular Ewald method in terms of the NDFT primitives. Then we show how the fast Fourier transform for nonequispaced (NFFT) can be applied, yielding an Ewald method based on the nonuniform fast Fourier transform.

Reciprocal Space Terms as DFT

We apply the generalized DFT, described in Section 3.1, to the calculation of the reciprocal-space energy and forces. This allows us to formulate the standard Ewald method for calculating the reciprocal energy and forces in terms of the NDFT primitives.

Reciprocal Energy

In the case of the electrostatic energy we have from Equation (28)

$$U_{qq}^{recip} = \frac{1}{2\pi L} \sum_{n \neq 0} \frac{e^{-\pi^2 |n|^2 /(\alpha L)^2}}{|n|^2} S(n) S(-n),$$

(46)

with the structure factor $S(n)$ defined as

$$S(n) = \sum_j q_j \exp\left(-\frac{2\pi \iota}{L} n \cdot r_j\right).$$

(47)

By comparing the definition of the transposed NDFT in Equation (45) and the structure factor in Equation (47) we note that they have the same structure; after a renumbering of the location indexes, the summation limits are also the same. In fact, by setting the normalized locations, $x_j = r_j/L$, and the samples, $\hat{f}_j = q_j$, we see by inspection that Equation (47) is a 3D instance of Equation (45) with $\hat{f}_n = S(n)$. Furthermore, assuming that the MD simulation box is centered around the origin, the normalized locations can be assumed to be in the domain $\mathcal{D}^d$ as defined in Equation (38).

Consequently, we can use the NDFT approach to calculate each of the components of the structure factor. From a computational point of view this means that we can also expect to utilize an NFFT based algorithm to calculate the components of the structure factor $S(n)$, rather than the straightforward summation normally used in the Ewald method.

Recasting Equation (46) in terms of Fourier-components

$$U_{qq}^{recip} = \frac{1}{2\pi L} \sum_{n \neq 0} \frac{e^{-\pi^2 |n|^2 /(\alpha L)^2}}{|n|^2} S(n) S(-n) = \frac{1}{\pi L} \sum_{n_z > 0} \frac{e^{-\pi^2 |n|^2 /(\alpha L)^2}}{|n|^2} \left|\hat{f}_n\right|^2.$$

(48)

Calculating the energy, U_{qq}^{recip}, using Equation (48) means that we

1)   calculate all $\hat{f}_n$ using the transposed NDFT,

2)   scale each $|\hat{f}_n|^2$ with a factor given by Equation (48),

3)   sum all the scaled components.

Reciprocal Forces

We calculate the contribution from the reciprocal-space forces using a similar approach as for the energy. In the formula below, $\text{Re}(\cdot)$, and $\text{Im}(\cdot)$, denote the real and imaginary part of the arguments, respectively. From Equation (24) we have that

$$F_i^{recip} = \frac{2q_i}{L^2} \sum_{n\neq 0} n \frac{e^{-\pi^2|n|^2/(\alpha L)^2}}{|n|^2} \left[\sin\left(\frac{2\pi}{L}n\cdot r_i\right)\text{Re}(S(n)) + \cos\left(\frac{2\pi}{L}n\cdot r_i\right)\text{Im}(S(n)) \right]. \tag{49}$$

Now, the structure factor $S(n)$ is just a complex number so the expression in the brackets in Equation (49) can be written as the imaginary part of a product

$$\sin\left(\frac{2\pi}{L}n\cdot r_i\right)\text{Re}(S(n)) + \cos\left(\frac{2\pi}{L}n\cdot r_i\right)\text{Im}(S(n)) = \text{Im}\left[\exp\left(\frac{2\pi\iota}{L}n\cdot r_i\right)S(n)\right]. \tag{50}$$

Inserting Equation (50) this into Equation (49) gives

$$F_i^{recip} = \frac{2q_i}{L^2} \sum_{n\neq 0} n \frac{e^{-\pi^2|n|^2/(\alpha L)^2}}{|n|^2} \text{Im}\left[\exp\left(\frac{2\pi\iota}{L}n\cdot r_i\right)S(n)\right]$$

$$= \text{Im}\left[\frac{2q_i}{L^2} \sum_{n\neq 0} n \frac{e^{-\pi^2|n|^2/(\alpha L)^2}}{|n|^2} \exp\left(\frac{2\pi\iota}{L}n\cdot r_i\right)S(n)\right]. \tag{51}$$

Note that Equation (51) is a vector equation. Furthermore, each of the three components has the same structure as the conjugated NDFT of Equation (44). By setting the normalized locations, $x_j = r_j/L$, and the samples,

$$\hat{g}_n = n \frac{e^{-\pi^2|n|^2/(\alpha L)^2}}{|n|^2} S(n) \quad \text{for } n \neq 0, \tag{52}$$

we see, again, by inspection that each component of Equation (51) is a 3D instance of Equation (44). Assuming that n is in the index set I_M of Equation (40), $n \in I_M$, and setting $g_0 = 0$, we can formulate F_i^{recip} directly in Fourier-terms

$$F_i^{recip} = \text{Im}\left[\frac{2q_i}{L^2}\sum_{n\in I_M}\hat{g}_n e^{2\pi in\cdot x_i}\right] = \frac{2q_i}{L^2}\text{Im}(g_i).$$

$$(53)$$

Calculating the reciprocal-space force F_i^{recip} on particle i, using Equation (53), means that we

1.   start with the structure factor components, $S(n)$, already obtained when we calulated U_{qq}^{recip},

2.   scale each $S(n)$ using Equation (52),

3.   giving a new set of Fourier-coefficients that are transformed back to real-space, via Equation (53), using the conjugated NDFT, and finally

4.   with Equation (53), taking the imaginary part of coefficient g_i and scaling it with $\dfrac{2q_i}{L^2}$ gives the reciprocal force on particle i.

Thus we can use the NDFT approach to calculate each of the components of the reciprocal forces. Again, from a computational point of view this means we can expect to utilize an NFFT based algorithm to find the respective components.

Combining the Ewald Method and NFFT

The reformulation of U_{qq}^{recip} and F_i^{recip} as in Equations (48) and (53) shows the central role of the transposed and conjugated NDFT in calculating the reciprocal-space energy and forces. Starting with the location of the charged particles, the structure factor $S(n)$ is calculated via a transposed NDFT. In the language of Ewald summation, we transform from real-space to reciprocal-space. Scaling the absolute value of the Fourier- components and summing gives U_{qq}^{recip}. To find F_i^{recip} we go back from the reciprocal-space to the real-space by first calculating the Fourier-components of the forces and then performing a conjugated NDFT.

An implementation of Ewald summation uses cut-offs, in reciprocal-space, n_{cut}, and real-space, r_{cut}; with α large enough and with a required accuracy, δ, truncate the sums Equation (27) and Equation (28) at the respective cut-offs so that the last term added $\leq \delta$, in each of the sums. When n_{cut} is fixed by

$$e^{-\pi^2|n|^2/(\alpha L)^2} \leq \delta \Rightarrow n_{cut} \geq \frac{\alpha L}{\pi}\sqrt{-\log(\delta)} \Rightarrow n_{cut} \propto L \propto N^{1/3}.$$

$$(54)$$

Then r_{cut} is determined by

$$\text{erfc}\left(\alpha r_{cut}\right) \approx e^{-\alpha^2 r_{cut}^2} \leq \delta \Rightarrow r_{cut} = \frac{\pi n_{cut}}{\alpha^2 L}.$$

(55)

We have recast U_{qq}^{recip} and F_i^{recip} in terms of Fourier-components and set $M_t = 2\sigma_s n_{cut}$, where σ_s is the oversampling factor. This gives $M_\Pi = \prod_{t=0}^{d-1} M_t$. In general the computational complexity of the NFFT method is $\mathcal{O}\left(M_\Pi \log M_\Pi + \log(1/\epsilon) N\right)$, where ϵ is the desired accuracy in the approximation used within NFFT [53].

Using Equation (54) and the above defintion of M_n, we see that the complexity becomes $\mathcal{O}\left(N \log N + \log(1/\epsilon) N\right)$. Note that ϵ is a function of m, for a fixed oversampling factor. With a controlled approximation of the structure factor via the use of nonuniform fast Fourier transform, the original computational complexity of $\mathcal{O}(N^{3/2})$ becomes $\mathcal{O}(N \log N)$.

At this stage, the path to a fast Ewald method should now be clear. By specifying an accuracy δ, we replace the transposed and conjugated NDFT with the corresponding operations using the NFFT algorithm. Thus Equations (48) and (53) become a concise procedure to calculate approximations of U_{qq}^{recip} and F_i^{recip}. Most of the mathematical details can be kept separate and hidden in a set of library routines and the remaining formulas pertain to the physics of the problem. Furthermore, with a library implementation based on a state-of-the-art FFT-library, we have good reason to expect it to be efficient.

Implementation and Results

In a first implementation [56,57], we used the libraries FFTW [58] and NFFT [54]. Details of the accuracy and scaling properties can be found in reference papers.

Basing the implementation on libraries has a number of advantages. It makes the implementation task easier and introduces a convenient division of labor in the program code: the mathematical aspects are mainly concentrated to the libraries while the physical aspects of the problem remain. Also, since the code becomes quite compact without becoming convoluted, it becomes easier to check, understand, and explain. Improvements and optimizations of the libraries can be easily included in the program, usually by just relinking the program. For example, the customization of the window function used in the NFFT algorithm---Gaussian functions, dilated cardinal B-splines, Sinc functions, or Kaiser-Bessel functions---is currently achieved by recompiling the NFFT library and relinking the application. Due to the comparatively small size of the NFFT library this is very quick. Furthermore, improvements in either theory or implementation of the used libraries will be easily accessible.

In summary, we claim that the ENUF method is

- efficient and concise, and
- has a clear separation of concerns between mathematical and physical details.

In a sense it can be said that we get the best of two worlds: a concise and efficient algorithm. The separation of mathematical concerns is a bonus that has the potential to simplify implementation and further developments due to the fact that they may occur independently of each-other.

ENUF IMPLEMENTED IN M.DYNAMIX

M.DynaMix is a highly modular general purpose parallel Molecular Dynamics code for simulations of arbitrary mixtures of either rigid or flexible molecules. It was released by Lyubartsev and Laaksonen in late 90's [59]. Most common force fields can be used in simulations with a variety of periodic boundary conditions (cubic, rectangular, hexagonal or a truncated octahedron). Quantum corrections to the atomic motion can be done using the Path Integral Molecular Dynamics approach. M.DynaMix has been used in applications from materials design to biological processes.

M.DynaMix deals with particle system interacting by a force field consisting of Lennard-Jones, electrostatic, covalent bonds, angles and torsion angles potentials as well as of some optional terms, in a periodic rectangular, hexagonal or truncated octahedron cell. Rigid bonds are constrained by the SHAKE algorithm [60]. In case of flexible molecular models the double time step algorithm is used [61]. Algorithms for NVE, NVT and NPT statistical ensembles are implemented, as well as Ewald sum approach for treatment of the electrostatic interactions. An option to calculate free energy by the expanded ensemble method with Wang-Landau optimization of the balancing factor is included in later versions. For its features and capabilities, M.DynaMix is diffused in a large modeling and simulation community. Written in FORTRAN 77, it can be run on a wide variety of hardware architectures both in sequential and parallel execution. The entire program source code consists of a number of FORTRAN files (modules), made of blocks of subroutines, performing different tasks or groups of tasks.

Framework Overview

The situation outlined for M.DynaMix is very common in the field of computational science: a package created in late 90's that is still at work in spite of new innovations in computing platforms. In a situation like this, each upgrade requires a trade-off between the need to preserve the existing structure and the wish to obtain as much efficiency as possible.

To take the best advantage from latest programming techniques and tools, we decide to use C language to create some of the new code segments. On one hand, modules in FORTRAN and C can easily coexist in the same code, just taking account of a few mild guidelines [62]; on the other hand, C language provides several enhancements related to intrinsic features (dynamic memory allocation and direct pointers reference) and possibility to set up complex data structures; furthermore, it allows a more plain access to a large amount of external routine libraries written in C. From a general point of view, M.DynaMix code has been projected with a good modularity degree and has been possible to make the most part of upgrades just switching a block (subroutine) with a new one.

Input Parameters

To use ENUF method for treating long range interactions, M.DynaMix user needs to set the proper key string in the input file [63]; parameters to specify are the Ewald convergence factor α and the number of points for FFT grid in x, y and z direction; starting from them, the program automatically sets proper values for over-sampling factor σ_s and approximation parameter p.

Implementation

Ewald-like methods for computing electrostatic energies typically replace part of the summation in real space with an equivalent summation in Fourier space; among M.DynaMix modules, there is one devoted to this reciprocal space duty. In the starting setup, this module was present in M.DynaMix in a version related to the full Ewald algorithm; ENUF implementation involved the creation of a second instance of this module.

A scheme of the ENUF module as implemented in M.DynaMix is presented in Table 1; the module in this case is a group of files, part of them written in FORTRAN and part in C; each file contains routines related to the algorithm steps. The table displays for each step the programming language used to write corresponding file; Some files make calls to external libraries, typically used to perform non uniform Fourier transform and its inverse.

Table 1: ENUF algorithm scheme: for each step is listed the language of the related routines and the presence of calls to external libs

Step	Language	Lib Calls
1: Coordinate scaling	F77	No
2: *Adjoint* FFT	C	Yes
3: Energy evaluation	F77	No
4: Factor rescaling	F77	No
5: *Regular* FFT	C	Yes
6: Forces evaluation	F77	No

Coordinate Scaling and Adjoint Transform

In the first step of the algorithm, particle coordinates are rescaled respect to the simulation box, in order to obtain the proper set of nonequispaced knots as required in Equation (38). Knots are stored in the unidimensional FORTRAN array X, in consecutive way, saving coordinates in reverse component order $\{z,y,x\}$ for each knot. Particle charges are located in unidimensional array Q.

At the next step, a C routine performs the 3-dim adjoint transform of charges Q located at knots X. C routine access to 1-dim FORTRAN vectors X and Q as three dimensional arrays; the reverse component order adopted in X vector filling is related to this and depends on the different way to arrange multi-dimensional arrays in memory: by columns for FORTRAN, by rows for C. Transformed data are stored in the complex array $\tilde{S}$.

Energy, Regular Transform and Forces

In this stage, energy is evaluated by a sum, according to formula (48). Then, the transformed array $\tilde{S}$ is rescaled using Equation (51); starting from $\tilde{s}$, three complex rescaled arrays are created, $\tilde{S}_x$, $\tilde{S}_y$ and $\tilde{s}_z$, one for each direction in the reciprocal space. A C routine back transform complex data set $\tilde{S}_x$, $\tilde{S}_y$ and $\tilde{S}_z$ related to the same set of knots X. In this step, three independent back transformations are performed, producing complex arrays F_x, F_y and F_z. Forces contributions for each particle are obtained according to Equation (53); components in x, y and z directions are evaluated using the imaginary part of arrays F_x, F_y and F_z.

External Libraries

To perform Fourier transform on nonequispaced data, our current implementation make use of an external existing function library. Among the available resources, we select the library NFFT [54]. NFFT is a widely diffused C subroutine library for computing nonequispaced discrete Fourier transform in one or more dimensions, of arbitrary input size and of complex data; it is based on FFTW [58].

Validation

In the original version of M.DynaMix the working method for the treatment of long range interactions is the full direct Ewald summation. Using a cross-check approach, namely comparing ENUF method to the stable one, it has been possible to debug every step of the algorithm implementation and check its correctness in deep way.

After the implementation, the same cross-check mechanism has been used to investigate ENUF precision and efficiency. Figure 1 displays the execution time spent for long range interactions evaluation respect to the number of particles N, when full Ewald and ENUF are used. In the simulation, of 100 iterations, the system is composed by 50 Na^+ ions and 50 Cl^- ions in water solution in a cubic cell. The number of water molecules is increased from 10^3 to 25×10^3 in seven discrete steps, keeping constant the density to 1.02 g/cm^3. The simulation run on a Intel Xeon workstation, 1.86 GHz, with ram 8 Gb.

The full Ewald method has been taken as reference; for each system size, ENUF parameters have been tuned to obtain the same Ewald precision, evaluated in terms of the electrostatics energy contribution, tolerating a maximum divergence of 2%.

The plot shows that execution time for long range interactions that represents the efficiency bottleneck in the original M.DynaMix version, has been drastically reduced saving the same numerical precision. The plot also shows the different trend for execution time vs N in the two methods; the point set for the full Ewald scheme is well fitted using the well-known theoretical behavior $\mathcal{O}(N^2)$; data points related to ENUF correspond quite well to the expected $\mathcal{O}(N \log N)$.

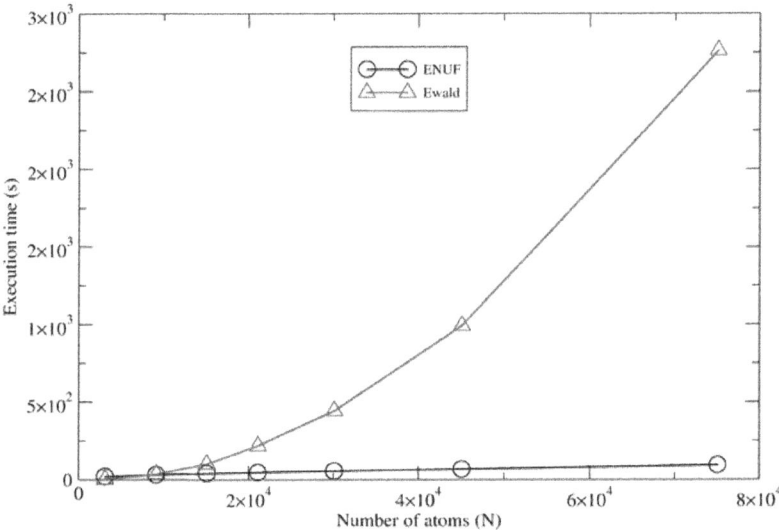

Figure 1. Execution time for long range interactions vs. number of particles N, for Ewald and ENUF. System is composed by 50 Na$^+$ and 50 Cl$^-$ in H$_2$O solution in a periodic cubic cell. Number of water molecules is increased from 10^3 to 25×10^3 keeping density to 1.02 g/cm^3.

Implementation of ENUF Method in Dissipative Particle Dynamics Scheme

The Dissipative Particle Dynamics (DPD) simulation method, originally proposed by Hoogerbrugge and Koelman, is a particle-based simulation method to simulate hydrodynamic phenomena at mesoscopic level [64, 65,66]. In DPD model, several atoms or molecules are grouped together to form coarse-grained particles. The interactions between any pair of DPD particles i and j are normally composed of three pairwise additive forces: the conservative force $F_{ij}^{C,DPD}$, the dissipative force $F_{ij}^{D,DPD}$, and the random force $F_{ij}^{R,DPD}$

$$f_{ij} = \sum_{i \neq j} \left(F_{ij}^{C,DPD} + F_{ij}^{D,DPD} + F_{ij}^{R,DPD} \right),$$

(56)

with

$$F_{ij}^{C,DPD} = \alpha_{ij} \omega^C \left(r_{ij} \right) \hat{r}_{ij},$$

(57)

$$F_{ij}^{D,DPD} = -\gamma \omega^D \left(r_{ij} \right) \left(v_{ij} \cdot \hat{r}_{ij} \right) \hat{r}_{ij},$$

(58)

$$F_{ij}^{R,DPD} = \sigma \omega^R \left(r_{ij} \right) \theta_{ij} \hat{r}_{ij},$$

(59)

where $r_{ij} = r_i - r_j$, $r_{ij} = |r_{ij}|$, $\hat{r}_{ij} = r_{ij}/r_{ij}$, and $v_{ij} = v_i - v_j$. The parameters α_{ij}, γ, and σ determine the strength of the conservative, dissipative, and random forces, respectively. θ_{ij} is a randomly fluctuating variable, with zero mean and unit variance.

The pairwise conservative force is written in terms of a weight function $\omega^C\left(r_{ij}\right)$, where $\omega^C\left(r_{ij}\right) = 1 - r_{ij}/R_c$ is usually chosen for $r_{ij} \leq R_c$ and $\omega^C\left(r_{ij}\right) = 0$ for $r_{ij} > R_c$ such that the conservative force is soft and repulsive. The unit of length R_c is related to the volume of DPD particles. Two weight functions $\omega^D\left(r_{ij}\right)$ and $\omega^R\left(r_{ij}\right)$ for dissipative and random forces, respectively, are coupled together through the fluctuation- dissipation theorem

$$\omega^D(r) = \left[\omega^R(r)\right]^2 \quad \text{and} \quad \sigma^2 = 2\gamma k_B T \tag{60}$$

to form a thermostat and generate naturally the canonical distribution (constant number of particles, N, volume, V, and temperature, T) [67]. In most applications, the weight function $\omega^D(r)$ adopts a simple form as [68]

$$\omega^D(r) = \left[\omega^R(r)\right]^2 = \begin{cases} \left(1 - \dfrac{r}{R_c}\right)^2 & (r \leq R_c) \\ 0 & (r > R_c) \end{cases}. \tag{61}$$

In the original DPD model, one critical advantage is the soft repulsive nature of conservative potential, which enables us to integrate the equations of motion with large time step. However, such advantage restricts the direct incorporation of electrostatic interactions in DPD model. The main problem is that dissipative particles carrying opposite point charges tend to collapse onto each other, forming artificial ionic clusters due to the stronger electrostatic interactions than soft repulsive conservative interactions.

In order to avoid such unphysical phenomena, point charges at the center of dissipative particles are usually replaced by charge density distributions meshed around particles to remove the divergency of electrostatic interactions at $r=0$ [69,70]. In our implementation [71,72], a Slater-type charge density distribution is considered with the form of

$$\rho_e(r) = \frac{q}{\pi \lambda_e^3} e^{\frac{-2r}{\lambda_e}}, \tag{62}$$

in which λ_e is the decay length. The integration of Equation (62) over the whole space gives the total charge q.

The electrostatic potential $\phi(r)$ generated by Slater-type charge distribution $\rho_e(r)$ can be obtained by solving Poisson's equation

$$\nabla^2 \phi(r) = -\frac{1}{\epsilon_0 \epsilon_r} \rho_e(r),$$

(63)

in which the variables ϵ_0 and ϵ_r are the permittivity of vacuum space and the dielectric constant of water at room temperature, respectively. In spherical coordinates, the Poisson's equation becomes

$$\frac{1}{r^2} \frac{\partial}{\partial r} \left(r^2 \frac{\partial}{\partial r} \phi(r) \right) = -\frac{1}{\epsilon_0 \epsilon_r} \rho_e(r).$$

(64)

By solving Poisson's equation, the electrostatic potential field $\phi(r)$ can be analytically expressed by

$$\phi(r) = \frac{1}{4\pi\epsilon_0 \epsilon_r} \frac{q}{r} \left(1 - \left(1 + \frac{r}{\lambda_e} \right) e^{\frac{-2r}{\lambda_e}} \right).$$

(65)

The electrostatic energy between two interacting charge density distributions i and j is the product of the charge density distribution i and the electrostatic potential generated by charge distribution j at position r_i

$$U_{qq}^{E,DPD} \left(r_{ij} \right) = q_i \phi_j \left(r_i \right) = \frac{1}{4\pi\epsilon_0 \epsilon_r} \frac{q_i q_j}{r_{ij}} \left(1 - \left(1 + \frac{r_{ij}}{\lambda_e} \right) e^{\frac{-2r_{ij}}{\lambda_e}} \right).$$

(66)

The electrostatic force on charge distribution i is the negative of the derivative of the potential energy $U_{qq}^{E,DPD}$ respect to its position r_i

$$F_i^{E,DPD} \left(r_{ij} \right) = -\nabla_i U_{qq}^{E,DPD} \left(r_{ij} \right)$$

$$= \frac{1}{4\pi\epsilon_0 \epsilon_r} \frac{q_i q_j}{\left(r_{ij} \right)^2} \left\{ 1 - e^{\frac{-2r_{ij}}{\lambda_e}} - \frac{2r_{ij}}{\lambda_e} e^{\frac{-2r_{ij}}{\lambda_e}} - \frac{2r_{ij}^2}{\lambda_e^2} e^{\frac{-2r_{ij}}{\lambda_e}} \right\}.$$

(67)

By defining parameter $r^* = r/R_c$ as the reduced center-to-center distance between two charge distributions and dimensionless parameter $\beta = R_c/\lambda_e$, respectively, the reduced electrostatic energy and force between two Slater-type charge distributions are given by

$$U_{qq}^{E,DPD}\left(r_{ij}^{*}\right)=\frac{1}{4\pi\epsilon_{0}\epsilon_{r}}\frac{q_{i}q_{j}}{\left(R_{c}r_{ij}^{*}\right)^{1}}\left\{1-\left(1+\beta r_{ij}^{*}\right)e^{-2\beta r_{ij}^{*}}\right\},$$

$$(68)$$

$$F_{i}^{E,DPD}\left(r_{ij}^{*}\right)=\frac{1}{4\pi\epsilon_{0}\epsilon_{r}}\frac{q_{i}q_{j}}{\left(R_{c}r_{ij}^{*}\right)^{2}}\left\{1-\left(1+2\beta r_{ij}^{*}\left(1+\beta r_{ij}^{*}\right)\right)e^{-2\beta r_{ij}^{*}}\right\}.$$

$$(69)$$

Comparing to the electrostatic energy and force between point charges in atomistic simulations in previous section, we can find that the electrostatic energy and force between two Slater-type charge distributions in DPD simulations are scaled with corresponding correction factors as

$$B_{U}=1-\left(1+\beta r^{*}\right)e^{-2\beta r^{*}},$$

$$(70)$$

$$B_{F}=1-\left(1+2\beta r^{*}\left(1+\beta r^{*}\right)\right)e^{-2\beta r^{*}}.$$

$$(71)$$

The similarities between electrostatic energy and force between point charges and counterparts between charge density distributions imply that once we get electrostatic energy and force between point charges, from which the electrostatic energy and force between Slater-type charge density distributions in DPD simulations can be directly rescaled with corresponding correction factors.

For the electrostatic energy and force between point charges, we know that both of them do diverge when the relative distance between point charges is close to 0. While for the electrostatic energy and force between charge density distributions, in the limit of $r_{ij}^{*}\rightarrow 0$, the reduced electrostatic energy and force between charge density distributions are, respectively, described by

$$\lim_{r_{ij}^{*}\rightarrow 0}U_{qq}^{E,DPD}\left(r_{ij}^{*}\right)=\frac{1}{4\pi\epsilon_{0}\epsilon_{r}}\frac{q_{i}q_{j}}{R_{c}}\beta,$$

$$(72)$$

$$\lim_{r_{ij}^{*}\rightarrow 0}F_{i}^{E,DPD}\left(r_{ij}^{*}\right)=0.$$

$$(73)$$

It is clear that the adoption of Slater-type charge distribution in DPD simulations removes the divergence of electrostatic interactions at $r_{ij}^{*}=0$, which means that both electrostatic energy and force between charge distributions are characterized with finite quantities.

By matching the maximum electrostatic energy between charge distributions at $r_{ij}^* = 0$ with Groot's previous work [69] gives $\beta = 1.125$. In our detailed implementations, we adopted a particular coarse-graining scheme [73] with $N_m = 4$ and $\rho = 4$, in which the former parameter means 4 water molecules being coarse-grained into one DPD particle and the latter means there are 4 DPD particles in the volume of R_c^3. With this particular scheme, the length unit R_c is given as $R_c = 3.107\sqrt[3]{\rho N_m} = 7.829\text{Å}$. From the relation of $\beta = R_c/\lambda_e$, we can get $\lambda_e = 6.954\,\text{Å}$, which is consistent with the electrostatic smearing radii used in González-Melchor's work [70].

Figure 2 shows the representation of reduced electrostatic energy and force with respect to the relative distance between Slater-type charge distributions. For a better comparison, we also include the typical soft conservative potential and force between dissipative particles, as well as the standard Coulombic potential and force between point charges, both of which do diverge at $r = 0$. In short range length scale, both the electrostatic energy and force are characterized with finite quantities, which attribute to the adoption of Slater-type charge distributions instead of point charges. In long range length scale, both the electrostatic energy and force are consistent with counterparts between point charges, which implys that the ENUF-DPD method can capture essential characteristics of electrostatic interactions at mesoscopic level.

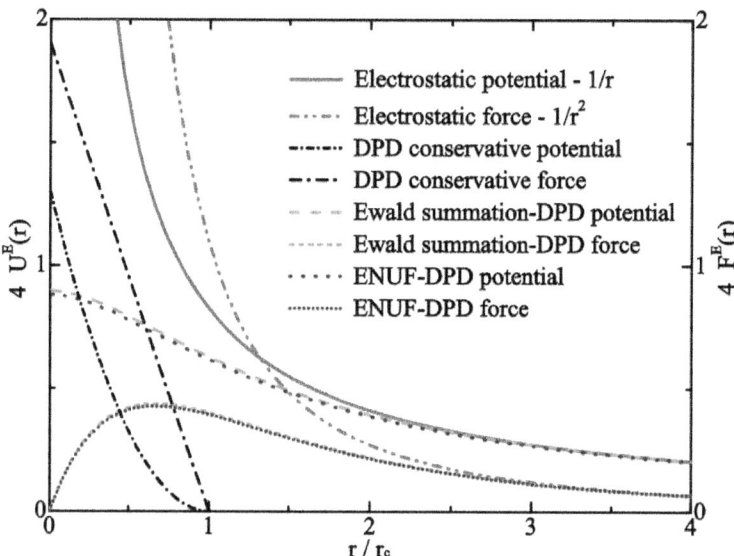

Figure 2: Electrostatic potential and force between two charge density distributions in DPD scheme calculated from the ENUF and Ewald summation methods with refer-

ence parameters. For a better comparison, the standard Coulombic potential and force, both of which diverge at $r=0$, and the typical conservative potential and force in standard DPD method are also included. Both the electrostatic potential and force expressions are plotted for two equal sign charge distributions.

Combining electrostatic force $F_i^{E,DPD}$ and soft repulsive conservative force $F_i^{C,DPD}$ gives the total conservative force F_i^{C*} on particles i in DPD simulations. The total conservative force F_i^{C*}, together with dissipative force $F_i^{D,DPD}$ and random force $F_i^{R,DPD}$, as well as the intramolecular bonding force $F_i^{S,DPD}$ for polymers and surfactants, act on dissipative particles and evolve the whole simulated system toward equilibrium conditions before taking statistical analysis.

As the number of charged DPD particles in the simulated system grows, the calculation of the reciprocal space electrostatic interactions will become the most time-consuming part. Using suitable parameters in

ENUF-DPD method assures that the time to calculate the real space summations is approximately the same as the time to calculate the reciprocal space summations, thereby reducing the total computational time. Herein we try to explore the ENUF-DPD related parameters and get a set of suitable parameters for further applications.

As addressed in Section 3.2, the implementation of ENUF-DPD method uses the Ewald convergence parameter α, required accuracy $\delta (\ll 1)$, and two cut-offs (r_{cut} for real space and n_{cut} for reciprocal space summations). These parameters are correlated with each other through two conditions shown in Equations (54) and (55). With required accuracy parameter $\delta = 1.0 \times 10^{-4}$, it is more convenient to pick a suitable value for n_{cut}. Then one can determine α and r_{cut} directly from Equations (54). However, due to the fact that n_{cut} should be integer and r_{cut} should be a suitable value for the cell-link list update scheme in DPD simulations, we adopt another procedure to get suitable parameters.

During the calculation of real space summation and self-interaction parts of electrostatic energy and force between point charges, we can directly multiply corresponding correction factors to get the electrostatic energy and force between slater-type charge density distributions. However, it is not accessible for the calculation of reciprocal space summation since we cannot rescale the NFFT transformation results. But if we choose suitable real space cutoff r_{cut}, beyond which two correction factors B_U and B_F converge to unit, one can directly adopt NFFT transformation result without any corrections. We find that electrostatic energy and force between charge density distributions are consistent with counterparts between point charges when the relative distance between two distributions is larger than $3.0R_c$. Hence in our simulations, $r_{cut} = 3.0R_c$ is taken as the cutoff for real space summations of electrostatic interactions.

By evaluating the Madelung constant of a face-centered cubic (FCC) lattice, we adopt the Ewald convergence parameter with the value of $\alpha = 0.20\,\text{Å}^{-1}$, which can generate accurate Madelung constant for FCC lattice structure and keep considerable accuracy. Then we perform coarse-grained simulations on bulk electrolyte system to explore suitable values for n_{cut}, and approximation parameter p in NFFT. It is specified that approximation parameter $p = 2$ and cutoff $n_{cut} = 7$ for reciprocal space summations can generate consistent electrostatic energy and force in comparison with those obtained from traditional Ewald summation method with reference parameters. Larger p values can further increase the accuracy of electrostatic interactions in ENUF-DPD method, but also the total computational time in treating electrostatic interactions increases. By compromising the accuracy and computational speed in the ENUF-DPD method, we adopt $p = 2$ and $n_{cut} = 7$ in following simulations.

With the set of explored optimized parameters, we address the computational complexity of ENUF-DPD method in treating electrostatic interactions. The computational complexity of ENUF-DPD method is appro- ximately described as $O(N \log N)$, which shows remarkably better computational efficiency than the tradi- tional Ewald summation method with acceptable accuracy in treating long-range electrostatic interactions between charged particles at mesoscopic level.

The ENUF-DPD method is then validated by investigating the influence of charge fraction of polyelectrolyte on corresponding conformational properties [71]. With the increase of charge fraction on polyelectrolyte, both the intramolecular correlations between charged beads on polyelectrolyte chain and the intermolecular correla- tions between charged beads on polyelectrolyte and counterions are enhanced. The conformation transition of polyelectrolyte chain from collapsed state to fully extended conformation can be visualized from simulations. Meanwhile, the dependence of the conformations of fully ionized polyelectrolyte on charge valency and concentration of added salts are also studied in details. Counterions with larger valency show stronger conden- sations on polyelectrolyte chains. Such counterions can induce polyelectrolyte chains from extended confor- mation to compact state, and then to swollen conformation with the increase of counterion concentrations.

With the ENUF-DPD method, we further investigate the specific binding structures of dendrimers on amphiphilic bilayer membranes [74].

We construct mutually consistent coarse-grained models for dendrimers and lipid molecules, which can properly describe the conformation of charged dendrimers and the surface tension of amphiphilic membranes, respectively. Systematic simulations are performed and simulation results reveal that the permeability of dendrimers across membranes is enhanced upon increasing dendrimer sizes. The negative curvature of amphiphilic membrane formed in dendrimer-membrane complexes is related to dendrimer concentration. Higher dendrimer concentration together with the synergistic effect between charged dendrimers can also enhance the permeability of dendrimers across amphiphilic membranes.

With these two typical and representative applications, we can see that the newly implemented ENUF-DPD method can capture the essential characteristics of electrostatic interactions at mesoscopic level. This method has all capabilities of ordinary DPD method, but includes applications where electrostatic interactions are essential but previously inaccessible, hence can be used to study charged complex systems at mesoscopic level.

SUMMARY AND CONCLUSIONS

Treatment of electrostatic interactions based on Ewald summation techniques is reviewed. While Ewald- summation is still considered as the most accurate scheme to compute the long-ranged interactions, it is also the part slowing down simulations. The scaling for large systems makes the computations very time-consuming. In particular, the reciprocal part of Ewald becomes a bottle-neck. As an attractive alternative approach to mesh- based schemes which show a linear scaling, we introduce an Ewald method based on non-uniform fast Fourier transforms (ENUF) giving examples of two implementations in already existing software packages, for ato- mistic Molecular Dynamics and Dissipative Particle Dynamics. We demonstrate that the implementation becomes straight-forward as we rely on NFFT library. We discuss the optimization of convergence parameters and window functions.

The ENUF method scales linearly as $\mathcal{O}(N \log N)$ and conserves both the energy and momentum to float point accuracy making it a very robust and accurate method.

ACKNOWLEDGEMENTS

Y.-L. W. and A. L. acknowledge the KA Wallenberg Foundation for financial support. F. M. acknowledges the Wenner-Gren and the Carl Tryggers Foundations for funding for Visiting Professorship. Y.-L. W. and A. L. thank the Kavli Institute for Theoretical Physics China (KITPC) and Chinese Academy of Sciences (CAS) for support and scientific environment during their stay. This work is supported by SERC the Swedish e-Science Center and Swedish Science Council. Parts of the computations were performed on resources provided by the Swedish National Infrastructure for Computing (SNIC) at PDC, HPC2N, and NSC.

REFERENCES

1.   1. M. P. Allen and D. J. Tildesley, "Computer Simulation of Liquids," Oxford University Press, New York, 1989.

2.   2. J. M. Haile, "Molecular Dynamics Simulation: Elementary Methods," John Wiley & Sons, New York, 1992.

3.   3. D. Frenkel and B. Smit, "Understanding Molecular Simulation: From Algorithms to Applications," Academic Press, Inc., Orlando, 1996.

4.   4. D. M. York, T. A. Darden and L. G. Pedersen, "The Effect of Long-Range Electrostatic Interactions in Simulations of Macromolecular Crystals: A Comparison of the Ewald and Truncated List Methods," Journal of Chemical Physics, Vol. 99, No. 10, 1993, p. 8345.http://dx.doi.org/10.1063/1.465608

5.   5. M. Bergdorf, C. Peter and P. H. Hünenberger, "Influence of Cut-off Truncation and Artificial Periodicity of Electrostatic Interactions in Molecular Simulations of Solvated Ions: A Continuum Electrostatics Study," Journal of Chemical Physics, Vol. 119, No. 17, 2003, p. 9129. http://dx.doi.org/10.1063/1.1614202

6.   6. C. Anézo, A. H. de Vries, H. D. Holtje, D. P. Tieleman and S. J. Marrink, "Methodological Issues in Lipid Bilayer Simulations," Journal of Physical Chemistry B, Vol. 107, No. 35, 2003, pp. 9424-9433. http://dx.doi.org/10.1021/jp0348981

7.   7. M. Patra, M. Karttunen, M. T. Hyvonen, E. Falck, P. Lindqvist and I. Vattulainen, "Molecular Dynamics Simulations of Lipid Bilayers: Major Artifacts Due to Truncating Electrostatic Interactions," Biophysical Journal, Vol. 84, No. 6, 2003, pp. 36363645.http://dx.doi.org/10.1016/S0006-3495(03)75094-2

8.   8. P. P. Ewald, "Die Berechnung Optischer und Elektrostatischer Gitterpotentiale," Annalen der Physik, Vol. 369, No. 3, 1921, pp. 253-287.http://dx.doi.org/10.1002/andp.19213690304

9.   9. S. W. de Leeuw, J. W. Perram and E. R. Smith, "Simulation of Electrostatic Systems in Periodic Boundary Conditions, I. Lattice Sums and Dielectric Constants," Proceedings of the Royal Society London,

Series A, Vol. 373, No. 1752, 1980, pp. 27-56.http://dx.doi.org/10.1098/rspa.1980.0135

10. 10. D. M. Heyes, "Electrostatic Potentials and Fields in Infinite Point Charge Lattices," Journal of Chemical Physics, Vol. 74, No. 3, 1981, pp. 1924-1929.

11. 11. T. Darden, D. York and L. Pedersen, "Particle Mesh Ewald: An NlogN Method for Ewald Sums in Large Systems," Journal of Chemical Physics, Vol. 98, No. 12, 1993, pp. 10089-10092. http://dx.doi.org/10.1063/1.464397

12. 12. D. York and W. Yang, "The Fast Fourier Poisson Method for Calculating Ewald Sums," Journal of Chemical Physics, Vol. 101, No. 4, 1994, pp. 3298-3300.http://dx.doi.org/10.1063/1.467576

13. 13. U. Essmann, L. Perera, M. L. Berkowitz, T. Darden, H. Lee and L. G. Pedersen, "A Smooth Particle Mesh Ewald Method," Journal of Chemical Physics, Vol. 103, No. 19, 1995, pp. 8577-8593. http://dx.doi.org/10.1063/1.470117

14. 14. P. F. Batcho and T. Schlick, "New Splitting Formulations for Lattice Summations," Journal of Chemical Physics, Vol. 115, No. 18, 2001, p. 8312.http://dx.doi.org/10.1063/1.1412247

15. 15. D. R. Wheeler and J. Newman, "A Less Expensive Ewald Lattice Sum," Chemical Physics Letters, Vol. 366, No. 5-6, 2002, pp. 537-543. http://dx.doi.org/10.1016/S0009-2614(02)01644-5

16. 16.Y.Shan,J.L.Klepeis,M.P.Eastwood,R.O.DrorandD.E.Shaw,"Gaussian Split Ewald: A Fast Ewald Mesh Method for Molecular Simulation," Journal of Chemical Physics, Vol. 122, No. 5, 2005, Article ID: 054101. http://dx.doi.org/10.1063/1.1839571

17. 17. P. J. Steinbach and B. R. Brooks, "New Spherical-Cutoff Methods for Long-Range Forces in Macromolecular Simulation," Journal of Computational Chemistry, Vol. 15, No. 7, 1994, pp. 667-683. http://dx.doi.org/10.1002/jcc.540150702

18. 18. C. L. Brooks III, B. M. Pettitt and M. Karplus, "Structural and Energetic Effects of Truncating Long Ranged Interactions in Ionic and Polar Fluids," Journal of Chemical Physics, Vol. 83, No. 11, 1985, p. 5897. http://dx.doi.org/10.1063/1.449621

19. 19. K. S. Kim, "On Effective Methods to Treat Solvent Effects in Macromolecular Mechanics and Simulations," Chemical Physics Letters, Vol. 156, No. 2-3, 1989, pp. 261-268. http://dx.doi.org/10.1016/S0009-2614(89)87131-3

20. 20. L. Greengard and V. Rokhlin, "A Fast Algorithm for Particle Simulations," Journal of Computational Physics, Vol. 73, No. 2, 1987, pp. 325-348.http://dx.doi.org/10.1016/0021-9991(87)90140-9

21. 21. J. E. Barnes and P. Hut, "A Hierarchical (NlogN) Force-Calculation Algorithm," Nature, Vol. 324, 1986, pp. 446-449. http://dx.doi.org/10.1038/324446a0

22. 22. J. Pérez-Jordá and W. Yang, "A Simple O(NlogN) Algorithm for the Rapid Evaluation of Particle-Particle Interactions," Chemical Physics Letters, Vol. 247, No. 4-6, 1995, pp. 484-490.

23. 23. Z. H. Duan and R. Krasny, "An Adaptive Treecode for Computing Nonbonded Potential Energy in Classical Molecular Systems," Journal of Computational Chemistry, Vol. 22, No. 2, 2001, pp. 184-195. 3.0.CO;2-7>http://dx.doi.org/10.1002/1096-987X(20010130)22:2<184::AID-JCC6>3.0.CO;2-7

24. 24. I. Tsukerman, "Efficient Computation of Long-Range Electromagnetic Interactions without Fourier Transforms," IEEE Transactions on Magnetics, Vol. 40, No. 4, 2004, pp. 2158-2160. http://dx.doi.org/10.1109/TMAG.2004.829022

25. 25. E. T. Ong, K. M. Lim, K. H. Lee and H. P. Lee, "A Fast Algorithm for Three-Dimensional Potential Fields Calculation: Fast Fourier Transform on Multipoles," Journal of Computational Physics, Vol. 192, No. 1, 2003, pp. 244-261. http://dx.doi.org/10.1016/j.jcp.2003.07.004

26. 26. C. Sagui and T. Darden, "Multigrid Methods for Classical Molecular Dynamics Simulations of Biomolecules," Journal of Chemical Physics, Vol. 114, No. 15, 2001, p. 6578. http://dx.doi.org/10.1063/1.1352646

27. 27. R. D. Skeel, I. Tezcan and D. J. Hardy, "Multiple Grid Methods for Classical Molecular Dynamics," Journal of Computational Chemistry, Vol. 23, No. 6, 2002, pp. 673-684. http://dx.doi.org/10.1002/jcc.10072

28. 28. J. A. Izaguirre, S. S. Hampton and T. Matthey, "Parallel Multigrid Summation for the N-Body Problem," Journal of Parallel and Distributed Computing, Vol. 65, No. 8, 2005, pp. 949-962. http://dx.doi.org/10.1016/j.jpdc.2005.03.006

29. 29. J. A. Baker and R. O. Watts, "Monte Carlo Studies of the Dielectric Properties of Water-Like Models," Molecular Physics, Vol. 26, No. 3, 1973, pp. 789-792.http://dx.doi.org/10.1080/00268977300102101

30. 30. I. G. Tironi, R. Sperb, P. E. Smith and W. F. van Gunsteren, "A Generalized Reaction Field Method for Molecular Dynamics Simulations," Journal of Chemical Physics, Vol. 102, No. 13, 1995, p. 5451. http://dx.doi.org/10.1063/1.469273

31. 31. T. N. Heinz and P. H. Hünenberger, "Combining the Lattice-Sum and Reaction-Field Approaches for Evaluating Long-Range Electrostatic Interactions in Molecular Simulations," Journal of Chemical Physics, Vol. 123, No. 3, 2005, Article ID: 034107. http://dx.doi.org/10.1063/1.1955525

32. 32. R. W. Hockney and J. W. Eastwood, "Computer Simulation Using Particles," McGraw-Hill, New York, 1981.

33. 33. B. A. Luty, M. E. Davis, I. G. Tironi and W. F. van Gunsteren, "A Comparison of Particle-Particle, Particle-Mesh and Ewald Methods for Calculating Electrostatic Interactions in Periodic Molecular Systems," Molecular Simulation, Vol. 14, No. 1, 1994, pp. 11-20. http://dx.doi.org/10.1080/08927029408022004

34. 34. X. Wu and B. R. Brooks, "Isotropic Periodic Sum: A Method for the Calculation of Long-Range Interactions," Journal of Chemical Physics, Vol. 122, No. 4, 2005, Article ID: 044107. http://dx.doi.org/10.1063/1.1836733

35. 35. T. M. Apostol, "Mathematical Analysis," Addison-Wesley, Boston, 1979.

36. 36. J. W. Perram, H. G. Petersen and S. W. de Leeuw, "An Algorithm for the Simulation of Condensed Matter Which Grows as the 3/2 Power of the Number of Particles," Molecular Physics, Vol. 65, No. 4, 1988, pp. 875-893. http://dx.doi.org/10.1080/00268978800101471

37. 37. J. Kolafa and J. W. Perram, "Cutoff Errors in the Ewald Summation Formulae for Point Charge Systems," Molecular Simulation, Vol. 9, No. 5, 1992, pp. 351-368. http://dx.doi.org/10.1080/08927029208049126

38. 38. S. W. de Leeuw, J. W. Perram and E. R. Smith, "Simulation of Electrostatic Systems in Periodic Boundary Conditions. II. Equivalence of Boundary Conditions," Proceedings of the Royal Society London, Series A, Vol. 373, No. 1752, 1980, pp. 57-66.http://dx.doi.org/10.1098/rspa.1980.0136

39. 39. S. W. de Leeuw, J. W. Perram and E. R. Smith, "Simulation of Electrostatic Systems in Periodic Boundary Conditions. III. Further Theory and Applications," Proceedings of the Royal Society London,

Series A, Vol. 388, No. 1794, 1983, pp. 177-193.http://dx.doi.org/10.1098/rspa.1983.0077

40.   40. W. H. Press and G. B. Rybicki, "Fast Algorithm for Spectral Analysis of Unevenly Sampled Data," Astrophysical Journal, Vol. 338, 1989, pp. 277-280.http://dx.doi.org/10.1086/167197

41.   41. J. W. Cooley and J. W. Tukey, "An Algorithm for the Machine Calculation of Complex Fourier Series," Mathematics of Computation, Vol. 19, 1965, pp. 297-301.http://dx.doi.org/10.1090/S0025-5718-1965-0178586-1

42.   42. A. Brandt, "Multilevel Computations of Integral Transforms and Particle Interactions with Oscillatory Kernels," Computer Physics Communications, Vol. 65, No. 1-3, 1991, pp. 24-38. http://dx.doi.org/10.1016/0010-4655(91)90151-A

43.   43. M. Pippig and D. Potts, "Parallel Three-Dimensional Nonequispaced Fast Fourier Transforms and Their Application to Particle Simulation," SIAM Journal on Scientific Computing, Vol. 35, No. 4, 2013, pp. 411-437. http://dx.doi.org/10.1137/120888478

44.   44. O. Ayala and L. P. Wang, "Parallel Implementation and Scalability Analysis of 3D Fast Fourier Transform Using 2D Domain Decomposition," Parallel Computing, Vol. 39, No. 1, 2013, pp. 58-77. http://dx.doi.org/10.1016/j.parco.2012.12.002

45.   45. Q. Liu and N. Nguyen, "An Accurate Algorithm for Nonuniform Fast Fourier Transforms (NUFFT's)," IEEE Microwave and Guided Wave Letters, Vol. 8, No. 1, 1998, pp. 18-20. http://dx.doi.org/10.1109/75.650975

46.   46. G. Steidl, "A Note on Fast Fourier Transforms for Nonequispaced Grids," Advances in Computational Mathematics, Vol. 9, No. 3-4, 1998, pp. 337-352.http://dx.doi.org/10.1023/A:1018901926283

47.   47. J. A. Fessler and B. P. Sutton, "Nonuniform Fast Fourier Transforms Using Min-Max Interpolation," IEEE Transactions on Signal Processing, Vol. 51, No. 2, 2003, pp. 560-574. http://dx.doi.org/10.1109/TSP.2002.807005

48.   48. C. Anderson and M. D. Dahleh, "Rapid Computation of the Discrete Fourier Transform," SIAM Journal on Scientific Computing, Vol. 17, No. 4, 1996, pp. 913-919.http://dx.doi.org/10.1137/0917059

49.   49. A. Duijndam and M. Schonewille, "Nonuniform Fast Fourier Transform," Geophysics, Vol. 64, No. 2, 1999, p. 539. http://dx.doi.org/10.1190/1.1444560

50. 50. J. P. Boyd, "A Fast Algorithm for Chebyshev, Fourier, and Sinc Interpolation onto an Irregular Grid," Journal of Computational Physics, Vol. 103, No. 2, 1992, pp. 243-257. http://dx.doi.org/10.1016/0021-9991(92)90399-J

51. 51. A. F. Ware, "Fast Approximate Fourier Transforms for Irregularly Spaced Data," SIAM Review, Vol. 40, No. 4, 1998, pp. 838-856.http://dx.doi.org/10.1137/S003614459731533X

52. 52. A. Nieslony and G. Steidl, "Approximate Factorizations of Fourier Matrices with Nonequispaced Knots," Linear Algebra and its Applications, Vol. 366, 2003, pp. 337-351. http://dx.doi.org/10.1016/S0024-3795(02)00496-2

53. 53. J. J. Benedetto and P. J. Ferreira, "Modern Sampling Theory: Mathematics and Applications," Springer, Berlin, 2001. http://dx.doi.org/10.1007/978-1-4612-0143-4

54. 54. S. Kunis and D. Potts, "NFFT3.2.3," Institute of Mathematics, University of Lübeck, D-23560 Lübeck, 2013. http://www.math.uni-luebeck.de/potts/nfft/

55. 55. F. Hedman and A. Laaksonen, "A Data-Parallel Molecular Dynamics Method for Liquids with Coulombic Interactions," Molecular Simulation, Vol. 14, No. 4-5, 1995, pp. 235-244. http://dx.doi.org/10.1080/08927029508022020

56. 56. F. Hedman, "Algorithms for Molecular Dynamics Simulations," Ph.D. Thesis, Stockholm University, Stockholm, 2006. http://www.diva-portal.org/smash/record.jsf?searchId=1&pid=diva2:178477

57. 57. F. Hedman and A. Laaksonen, "Ewald Summation Based on Nonuniform Fast Fourier Transform," Chemical Physics Letters, Vol. 425, No. 1-3, 2006, pp. 142-147. http://dx.doi.org/10.1016/j.cplett.2006.04.106

58. 58. M. Frigo and S. G. Johnson, "FFTW-The Design and Implementation of FFTW3," Proceedings of the IEEE, Vol. 93, No. 2, 2005, pp. 216-231. http://dx.doi.org/10.1109/JPROC.2004.840301

59. 59. A. P. Lyubartsev and A. Laaksonen, "M.DynaMix—A Scalable Portable Parallel MD Simulation Package for Arbitrary Molecular Mixtures," Computer Physics Communications, Vol. 128, No. 3, 2000, pp. 565-589. http://dx.doi.org/10.1016/S0010-4655(99)00529-9

60. 60. J. P. Ryckaert, G. Ciccotti and H. J. C. Berendsen, "Numerical Integration of the Cartesian Equations of Motion of a System with Constraints: Molecular Dynamics of n-Alkanes," Journal

of Computational Physics, Vol. 23, No. 3, 1977, pp. 327-341. http://dx.doi.org/10.1016/0021-9991(77)90098-5

61.   61. M. Tuckerman, B. J. Berne and G. J. Martyna, "Reversible Multiple Time Scale Molecular Dynamics," Journal of Chemical Physics, Vol. 97, 1992, pp. 1990-2001.

62.   62.   http://www.yolinux.com/TUTORIALS/ LinuxTutorialMixingFortranAndC.html

63.   63. A. P. Lyubartsev, M. DynaMix User Manual.http://www.fos.su.se/ sasha/md_prog.html

64.   64. P. J. Hoogerbrugge and V. A. Koelman, "Simulating Microscopic Hydrodynamic Phenomena with Dissipative Particle Dynamics," Europhysics Letters, Vol. 19, No. 3, 1992, p. 155. http://dx.doi.org/10.1209/0295-5075/19/3/001

65.   65. V. A. Koelman and P. J. Hoogerbrugge, "Dynamic Simulations of Hard-Sphere Suspensions under Steady Shear," Europhysics Letters, Vol. 21, No. 3, 1993, p. 363.http://dx.doi.org/10.1209/0295-5075/21/3/018

66.   66. Z. Y. Lu and Y. L. Wang, "An Introduction to Dissipative Particle Dynamics," In: L. Monticelli and E. Salonen, Eds., Methods in Molecular Biology: Biomolecular Simulations, Humana Press, New York, 2013, pp. 617-633. http://link.springer.com/protocol/10.1007/978-1-62703-017-5_24

67.   67. P. Espanol and P. Warren, "Statistical Mechanics of Dissipative Particle Dynamics," Europhysics Letters, Vol. 30, No. 4, 1995, p. 191. http://dx.doi.org/10.1209/0295-5075/30/4/001

68.   68. R. D. Groot and P. B. Warren, "Dissipative Particle Dynamics: Bridging the Gap between Atomistic and Mesoscopic Simulation," Journal of Chemical Physics, Vol. 107, No. 11, 1997, p. 4423. http://dx. doi.org/10.1063/1.474784

69.   69. R. D. Groot, "Electrostatic Interactions in Dissipative Particle Dynamics-Simulation of Polyelectrolytes and Anionic Surfactants," Journal of Chemical Physics, Vol. 118, No. 24, 2003, p. 11265. http:// dx.doi.org/10.1063/1.1574800

70.   70. M. González-Melchor, E. Mayoral, M. E. Velázquez and J. Alejandre, "Electrostatic Interactions in Dissipative Particle Dynamics Using the Ewald Sums," Journal of Chemical Physics, Vol. 125, No. 22, 2006, p. 224107. http://dx.doi.org/10.1063/1.2400223

71.   71. Y. L. Wang, A. Laaksonen and Z. Y. Lu, "Implementation of Non-Uniform FFT Based Ewald Summation in Dissipative Particle Dynamics

Method," Journal of Computational Physics, Vol. 235, 2013, pp. 666-682. http://dx.doi.org/10.1016/j.jcp.2012.09.023

72. 72. Y. L. Wang, "Electrostatic Interactions in Coarse-Grained Simulations: Implementations and Applications," Ph.D. Thesis, Stockholm University, Stockholm, 2013.http://www.diva-portal.org/smash/record. jsf?pid=diva2:641063

73. 73. A. AlSunaidi, W. K. den Otter and J. H. R. Clarke, "Liquid-Crystalline Ordering in Rod-Coil Diblock Copolymers Studied by Mesoscale Simulations," Philosophical Transactions of the Royal Society London A, Vol. 362, No. 1821, 2004, pp. 1773-1781.http://dx.doi.org/10.1098/rsta.2004.1414

74. 74. Y. L. Wang, Z. Y. Lu and A. Laaksonen, "Specific Binding Structures of Dendrimers on Lipid Bilayer Membranes," Physical Chemistry Chemical Physics, Vol. 14, No. 23, 2012, pp. 8348-8359. http://dx.doi. org/10.1039/c2cp40700k

Chapter 7

COMPUTER SIMULATION OF RADIATION DEFECTS IN GRAPHENE AND RELATIVE STRUCTURES

Arkady M. Ilyin[1]

[1]Kazakh National University, Physical Department, Kazakhstan

INTRODUCTION

Graphene is a single layer of carbon atoms arranged in a chicken-wire-like hexagonal lattice and in spite of its recent availability for experimental investigations (Novoselov et al., 2004) it is an object of great interest for many researchers (Elias et al., 2009; Sofo et al., 2007; Luo et al., 2009; Teweldebrhan and Balandin, 2009; Ilyin et al., 2009) because of amazingly wide field of its potential applicability: electronics, sensors, materials science, biology etc. In particular, graphene and few layer graphene fragments as well as carbon nanotubes can be used in production of composites, based on metal, ceramic, polymer matrices, filled with graphene's or few layer graphene fragments as elements of reinforcement. Obviously, the main goal of using graphene or nanotubes in making composites is using their extremely high mechanical characteristics in combination with low weight. It should be noticed, that many difficulties concerning graphene's applications originate from its surface chemical inertness. In other words, the sp^2 electron structure of ideal graphene often results in very low binding energy between graphene's surface and atoms of many elements. It is one of the obstacles for modifying and applications of graphene in production of electronic devices, when controllable electronic properties are needed. Besides, it results in poor interfacial bonding of the graphene fragments with matrices in composite materials and with sliding between few layer graphene sheets under stressed state. The noticeable success in this direction has been recently achieved in the work (Elias et al., 2009), in which the hydrogenation of graphene was performed by using an ion-plasma technique. It should be noticed, that composite of graphene and hydrogen, associated with graphene's surface was theoretically predicted few years ago in the paper (Sofo et al., 2007) and named as graphane. Unfortunately, this success up to now can be surely referred only to graphene-hydrogen composition. But today's technologies especially in the field of materials science need much more wide area of possible compositions and special materials. It was reasonable to suppose, that radiation defects may essentially improve binding ability of graphene with atoms of other elements due to production of additional chemical bonds (Ilyin, 2010 ; Ilyin & Beall, 2010). It is especially important for application of graphene species in R&D of composites. Moreover defects in such structures may improve mechanical properties by linking reinforcing carbon nanoelements to each other and by increasing the strength and stiffness of the composite (Ilyin et al.,2010). Unfortunately, it is not yet well understood which kinds of stable radiation defects and their complexes can exist in graphene and its derivatives. Obviously, it is not easy to create definite types of radiation defects and perform direct studies of them in nanoobjects in direct laboratory experiments. In this situation computer simulation of radiation defects in graphene- and relative nanostructures becomes of great importance (Ilyin et al., 2011). In this paper we have submitted some results of computer simulation and calculations of some possible kinds of stable radiation defects in graphene, and more complex configurations, involving atoms of light metals: Be and Al, which are linking with radiation defects. In other words, we consider possible effect of radiation defects on production and modification of graphene – metal composites and calculate energetic and structural properties for some mostly possible configurations. Computer simulation and calculations were performed by the use well known molecular dynamics, extended Hückel technique and density functional theory.

SIMULATION AND CALCULATIONS

The first model which we consider in this chapter is a simple configuration, which consists of a graphene sheet and a carbon atom adsorbed with it. We imply, that in this case no chemical bond between graphene and C atom exists. This kind of interaction refers to well known van der Waals forces. These interactions are weak and insignificant in our common life, but their role increases dramatically among the nano-scaled objects. One of the important features of them - the additivity i.e. every particle of the system makes its contribution into the total interaction energy. Therefore, they can be relatively large for nanosystems, involving $10^2 - 10^3$ atoms or molecules. But van der Waals interactions are not accounted by the use widespread computer techniques like molecular orbital linear combination of atomic orbitals by self consistent field (MO LCAO SCF) or density functional theory (DFT). Therefore, in this problem the method of molecular dynamics (MD), which is recognized as the effective tool for similar systems, was chosen. A graphene sheet for MD simulation of single- and diatomic defects on the undamaged structure was built of 78 atoms. Van der Waals interaction between adsorbed atoms and graphene was described by well known Lennard-Johns potential in the usual form.

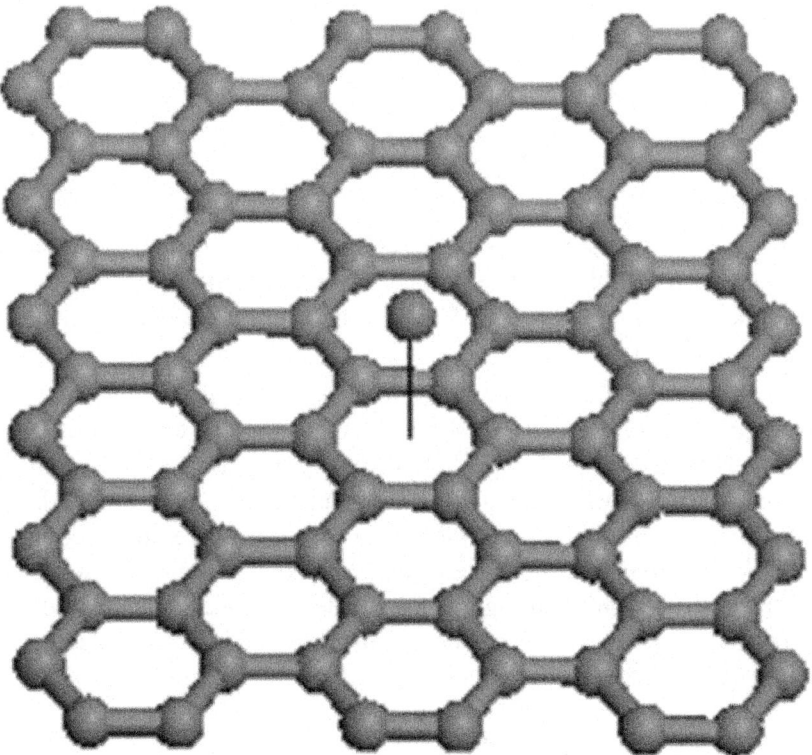

Figure 1: Configuration of a single adsorbed atom on a graphene structure.

Presented in Figure 1 is one of investigated defect configurations which is actually a single carbon atom, adsorbed on the surface of the undamaged graphene sheet. The maximum value of the binding energy for the single adsorbed atom $E = -0.18$ eV at a distance $Z = 0.25$ nm from the graphene plane. One important characteristic is also the energy of carbon atom in the center point of graphene's hexagon ($Z = 0$). This position was found to be very unstable with the positive energy equals to $E_0 = 11.4$ eV. It means also, that the graphene sheet is impermeable for displaced carbon atoms with energies lower than E_0.

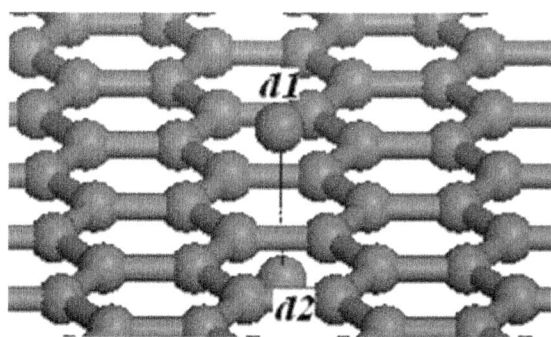

FIGURE 2: Configuration of a dumbbell defect on a graphene (N=78) structure.

Figure 2 presents a configuration of a more complicated, two-atom defect which is like a dumbbell with a symmetrical configuration of the atoms d1 and d2 normal to the graphene sheet. Calculations of this defect were performed by MD, using the LJ potential.

To begin, in all cases the minimum energy lateral position of the atoms adsorbed, had been found out at the normal axis of symmetry of hexagon (Z-axis). Further we performed calculations with movement of atoms along the Z-axis. Results of calculations of the binding energies for these defects as a function of a distance Z over the center of hexagon are presented in Figure 3. It can be seen, that there is an interval between approximately 2 and 3 angstroms that exhibits a trough with a negative energies, which is evidence of the existence of stable binding states. Low values of bonding energies testify of the vdW nature of the interaction.

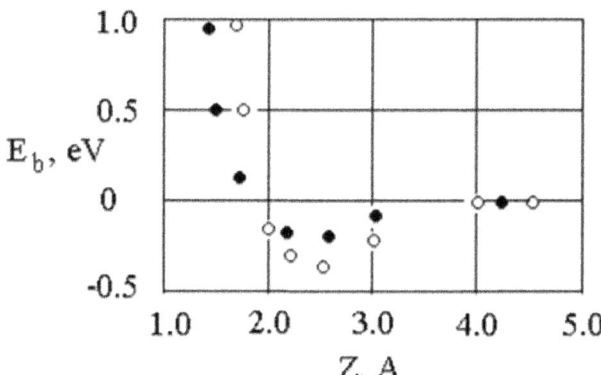

Figure 3: The binding energy for the single atom and the dumbbell defect as a function of a distance Z over the center of hexagon. The black marks – the single atom configuration (see Figure 1), the light marks – the dumbbell configuration (see Figure 2)

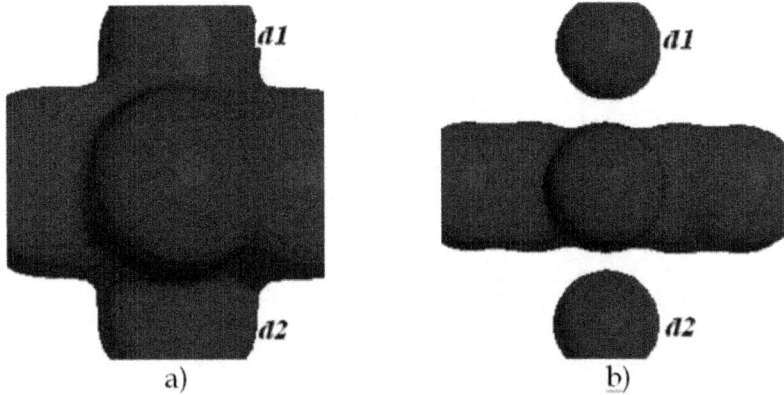

Figure 4: The electron charge distribution for the dumbbell presented in Fig.2. Density of electron charge equals: a) 0.02 el / $\mathring{A}^3$, b) 0.5 el / $\mathring{A}^3$.

Figure 4 illustrates the results of calculations of electron charge distribution for the dumbbell presented in Figure 2, performed by MO LCAO method. These calculations were performed in order to check our assumption about vdW interaction between atoms adsorbed and graphene. One can see some overlapping of electron charge only by very low level of electron density (Figure 4,a). And, obviously, there are no signs of overlapping of electron charge at high level of electron density, which could be responsible for some kind of bonding between graphene and atoms adsorbed. One can see well distinguished electron charge clouds, obviously closed on graphene and d1 and d2 atoms with a gap between them. At the same time graphene's structure is linked with dense electron clouds, which provided strong bonding. It proves that weak bonding interaction for the defects presented above is controlled by vdW forces. It is unlikely, that such defects can be useful for essential modifying of mechanical properties of composite materials.

Vacancies in Graphene

Irradiation of graphene-based electronic devices by fast electrons or ions will be always accompanied by creation of atom vacancies. Therefore, it is very important to know about changes in electronic properties of graphene fragments which should be expected under irradiation and about how they depend on defect concentration. For such estimation we have used large enough graphene's fragment (N=208).

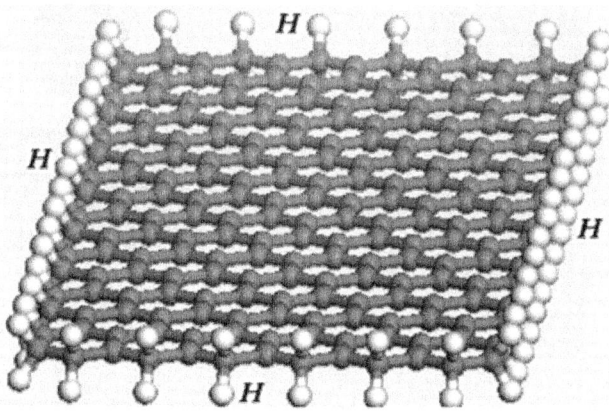

Figure 5: Graphene for calculations with edge-bonds shut by H-atoms.

In order to avoid end-effects by calculations, free end-bonds of carbon atoms were shut with hydrogen atoms. Afterwards, in order to take into account the possible effect of a larger size of a real graphene sheet, which can restrict the atoms neighboring to the vacancy, all edge atoms of graphene were fixed at their initial positions. After that we simulated and calculated one-, two- and three –vacancy configurations with using in all cases a procedure of energy minimization.

It was revealed, that in all cases, after energy optimization the vacancy zone increased so that all the three two-coordinated atoms, neighboring the vacancy, were shifted nearly symmetrically: all three distances between surrounding atoms 1-2, 2-3, 3-1 (Figure 6) become as large as 2.76 Å instead of 2.46 Å in the initial state.

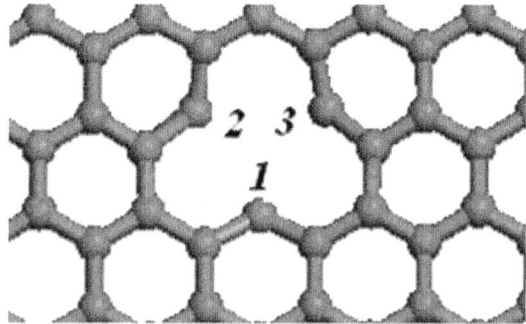

Figure 6: Vacancy zone in graphene after the optimization procedure.

Figure 7 presents a configuration of a graphene with 3 single vacancies displayed by the electron charge distribution.

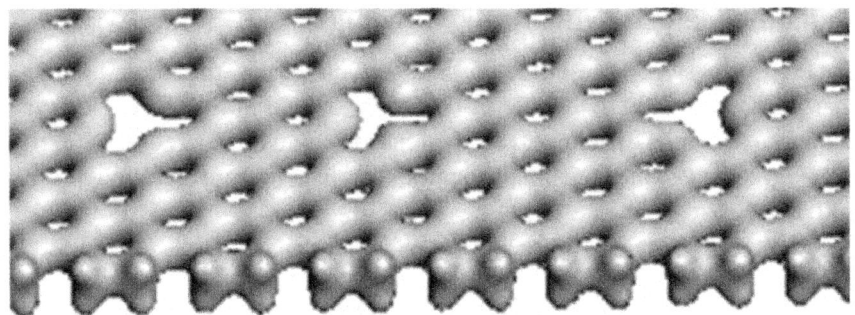

Figure 7: Graphene with 3 single separated vacancies in the structure.

Calculations of HOMO and LUMO were also performed in all cases. Figure 8 presents dependence of Eg = HOMO − LUMO for graphene -208 with different numbers of vacancies. There were one-, two-, and three vacancies in aligned configuration.

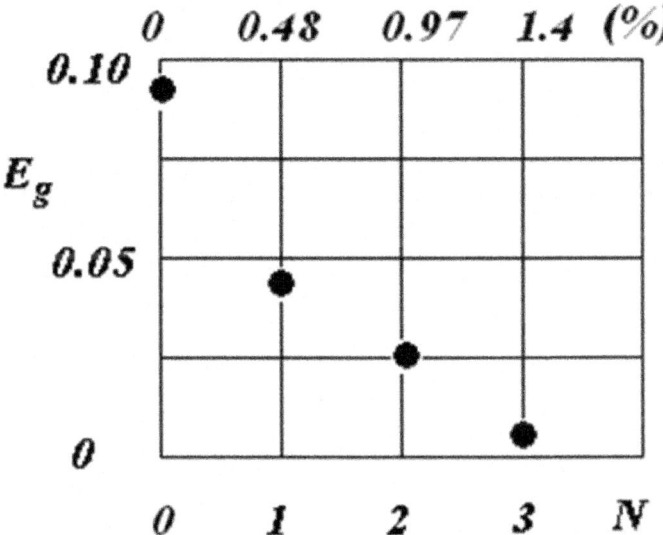

Figure 8: The value of Eg = HOMO − LUMO as a function of concentration of vacancies in a graphene-208 fragment.

Obviously, data for E_g presented in Figure 8 illustrate the effect of limited size of graphene fragment (so called size-effect), because the infinite graphene (the initial state with N =0) is intrinsically semi-metal with Eg =0. But these and similar effects should be taken into account when physicists and technologists will design devices based on real graphene fragments of limited sizes.

Radiation Defects With Strong Bonding

As the next step we simulated and calculated energetic and structural characteristics of 3D defect configuration presented in Figure 9. This type of radiation defect, which involves two carbon atoms arranged symmetrically over a vacancy can be named "dumbbell", like to configuration presented inFigure 2.

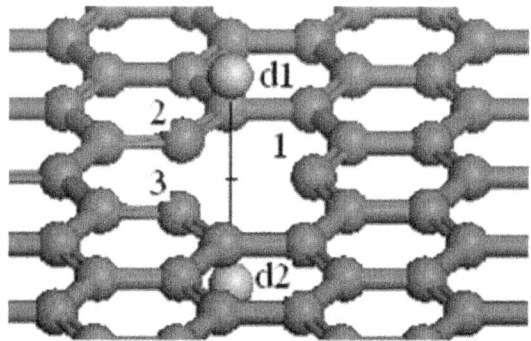

Figure 9: The complex radiation defect, involving a vacancy and a dumbbell configuration.

But in this case the two carbon atoms (d1,d2) of the dumbbell are chemically bonded with free bonds of atoms, neighboring at the vacancy. One can see from the graph in Figure 10, that there is a strong bonding.

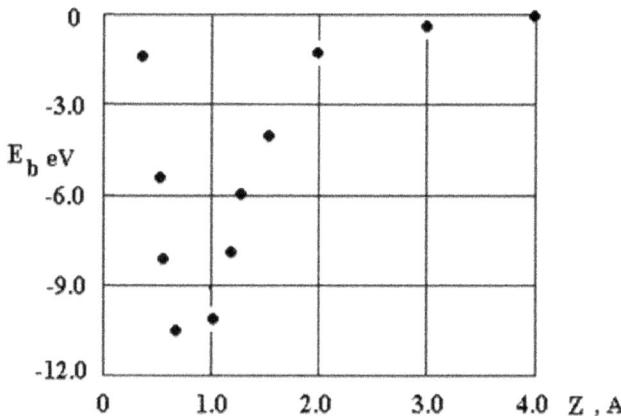

Figure 10: The binding energy for the dumbbell placed over a vacancy as a function of the distance Z over the graphene sheet.

Obviously, the elastic properties of composite materials and their uniformity are of great importance in using composite materials. The E_b - Z curve in Figure 10 can be used for estimation of elastic characteristics of the C-C dumbbell defect along the graphene sheet (under shear stress). One can see, that the maximum slope of the curve is near the point Z = 1.3 Å. The numerical estimation by using

$$\Delta E / \Delta Z \Delta E / \Delta Z$$

at this point with small segments gives the value 0.5 TPa. The maximum binding energy of the dumbbell over the relaxed vacancy was obtained as large as -10.0 eV and the corresponding distance between the graphene's plate and atoms equals 0.7 A. The electron charge distribution presented in Figures 11 and 12 proves that there is fast chemical bond between dumbbell and graphene sheet.

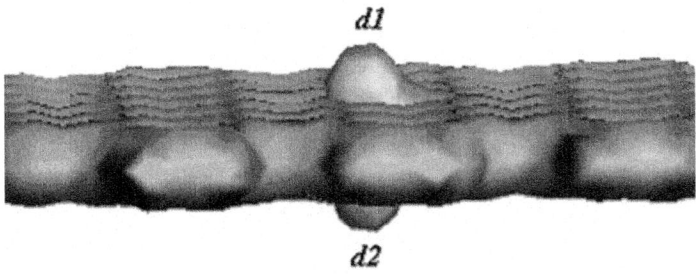

Figure 11: The electron charge distribution for the graphene-dumbbell configuration at density of the charge equal to 1.0 el/Å³.

Figure 12: The electron distribution for a graphene with two separated dumbbells.

Calculation of electronic properties of dumbbell configurations presented in Figures 11 and 12 for graphene-208 were also performed. Eg = HOMO − LUMO was as large as 0.05 eV for a case with one- and two dumbbels. Figure 12 displayed no signs of noticed non-uniformity between two dumbbells in the graphene structure. So that, effect of radiation defects like dumbbells can be considered as insignificant by their concentrations about 1%. The large value of the binding energy for the dumbbell defects and the electron charge distribution calculated for these defect configurations obviously demonstrate, that there is significant interaction between the dumbbell and a graphene, as well as between atoms of the dumbbell itself.

The ability of graphene's vacancy to bind atoms of other elements by the chemical way were firstly checked by simulation and calculation of the vacancy - hydrogen complex defect, which was performed in the hydrogenated graphene ("chair"- graphane) structure (Figure 13). Our calculations proved that a vacancy zone can serve as a site with a high concentration of hydrogen. In this case the vacancy zone has non- symmetrical mode of deformation: distances between atoms: C1-C2 = C2-C3 = 2.72 Å, C1-C3 = 2.55 Å

The results of calculations of binding energy of H atoms, bonding at the vacancy are presented in Fig. 14. These data witness, that the value of binding energy depends on the total number of H atoms, placed at the vacancy. One can see, that the binding energy has well defined minimum at N = 3.

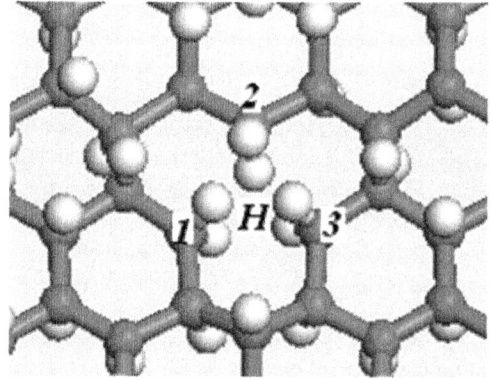

Figure 13: The atomic structure of vacancy with 6 hydrogen atoms bonding.

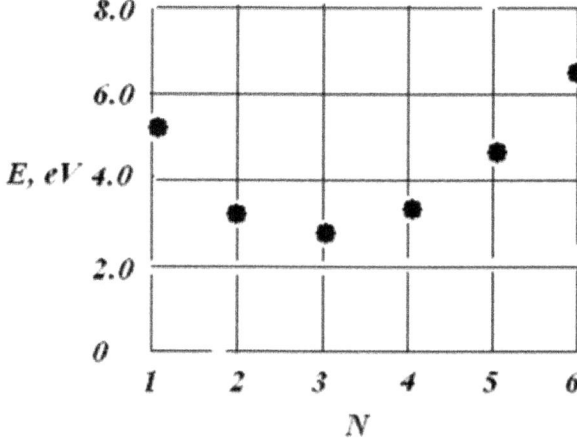

Figure 14: The dependence of hydrogen – vacancy binding energy on the total number N of hydrogen atoms linking with the vacancy.

Bridge-Like Radiation Defects In Graphene

It is known very well that two layer graphene is being one of main components by production of graphene's materials by many technologies. Therefore, it can play a significant role in some applications, using graphene materials. At least, it is reasonable to study properties of two layer graphene, keeping in mind its future applications. In particular, two layer graphene fragments can be used in production of lightweight composite materials with high stiffness and strength. Therefore, calculations were made also for bilayer graphene fragment. The usual AB graphite-like configuration, which is the most common for graphite –like materials, was chosen for calculations (see Figure 15). The Figure 16 illustrates that interaction between undamaged graphene sheets has van der Waals nature, without any signs of electron charge overlapping. The coupled atom pair which were removed by creating the vacancy pair is marked by black. The interstitial C-atom, knocked from the structure was placed between graphene layers. After that relaxation procedure was used to obtain a minimum of the total energy of the defect volume. The edge atoms of graphenes were fixed in order accounting the size effect.

In Figure 16 one can see the typical picture of the electron charge distribution for undamaged bilayer graphene, controlled by van-der-Waals interactions. There is no electron charge overlapping, between different graphene sheets.

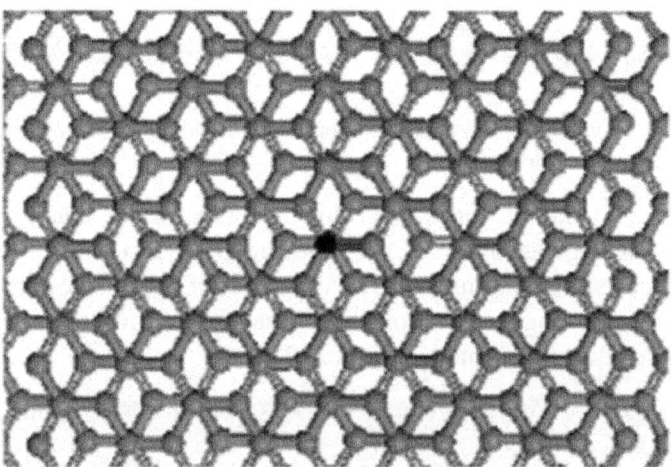

Figure 15: Graphene AB two layer used for building complex interior defect.

In Figure 17 one can see the much more complex defect configuration with vacancies, faced each other, which were produced in both of graphene sheets and interstitial carbon atom (*i*), caught between them. This type of radiation defects can be called as a bridge-like defect. The essential feature of the defect is that the two graphene sheets are linked with fast covalent bond based on the interstitial atom and as one can see, neighboring atoms 1-2 and 3-4 facilitating the rising two additional bonds, because of pulling in the gap between graphenes. The distribution of the electron charge presented inFigure 18 proves existing of three covalent bonds, originated between graphene layers.

Figure 16: The distribution of the electron charge for undamaged bilayer graphene (the density equals 0.4 el / A^3).

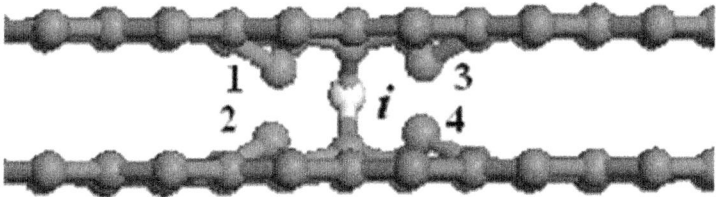

Figure 17: A complex bridge-like radiation defect bonding together sheets of bilayer graphene after relaxation.

The total binding energy for this complex defect configuration was calculated as large as -11.3 eV. We have supposed, that ends of two- or few layer graphene fragments may also serve as sites of bridge-like defects, linking graphene's sheets together. A typical configuration of end-bridge-like bonding of a carbon atom is presented in Figure 19.

Figure 18: The distribution of the electron charge (the density equals 1.4 el / A^3).

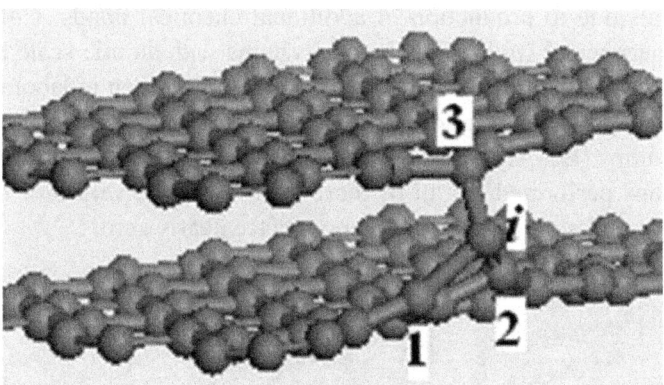

Figure 19: The bridge-like defect based on an interstitial atom *i* at the edge of bilayer graphene.

GRAPHENE-METAL COMPOSITION INDUCED BY RADIATION

It is known very well, that graphene and few layer graphene fragments, like to carbon nanotubes can be used as elements of reinforcement in production of composites, based on various matrices. Moreover, many important physical properties of material (particularly, electric and heat conduction, magnetic characteristics and so on) can be modified and improved by using graphene and few layer graphene fragments as filler. As was mentioned above, many difficulties concerning graphene's applications originate from its sp² electron structure. In other words, the electron structure of ideal graphene often results in very low binding energy between graphene's surface and atoms of many metals. It is one of the obstacles for modifying and applications of graphene in production of composite materials. In our recent papers we suggested using of radiation modification of composite materials with carbon nanostructures due irradiation by fast electrons or ions. Production of special kind of bridge-like defects may be considered as an effective technological tool of essential modification of physical-mechanical properties of composite materials, filled with carbon nanostructures (Ilyin & Beall, 2010).

Beryllium, aluminum and their alloys are being very important materials for designing new composites, especially for fields where combination of light weight with high strength is needed, for example – transportation systems and air-space technologies. Therefore, in this paper we focused on study of possible production composite materials based on Be and Al matrices, with using graphene fragments as reinforcement elements. We suppose, that radiation defects may essentially improve binding ability of graphene with atoms of light metals due to production of additional chemical bonds. Unfortunately, direct experimental study of such nanosystems with atomic scale defects and operations with them can be hardly performed today even in laboratories with high level equipment. Figure 20presents some typical possible positions of metallic atoms (Be, Al) arranged on graphene surface in high symmetry sites. Calculations performed for all of these positions gave values of the binding energy of metal atoms on the graphene surface nearly zero.

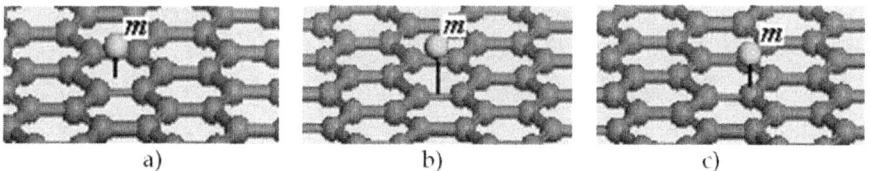

a)   b)   c)

Figure 20: Some of calculated symmetrical configurations of metal atoms (Be and Al) on graphene surface with nearly zero binding energies: a) over the center of a hexagon; b) over the center of the C-C bond; c) over a C atom.

BE – Graphene Composition

Figure 22 presents a scheme of estimation elastic characteristics of defect involving Be dumbbell at vacancy by techniques like above for C-C dumbbell. The value of elastic modulus for direction "to right" in the Figure 21 was calculated as large as 0.05 TPa and 0.02 TPa in the perpendicular direction.

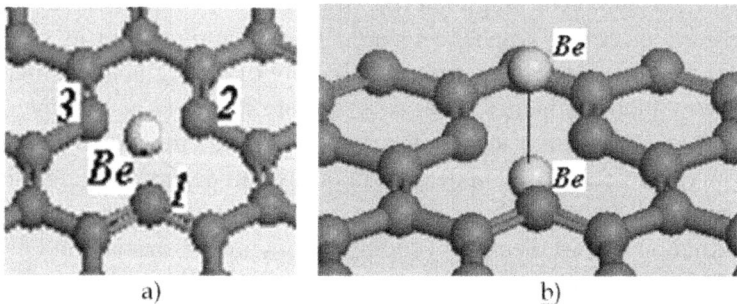

a)   b)

Figure 21: Configurations of Be atoms bonded with a vacancy: a) a stable position of the single Be atom in the graphene sheet. The binding energy E_b equals 2.6 eV; b) Configuration of "Be – dumbbell" over a vacancy with binding energy 4.1 eV.

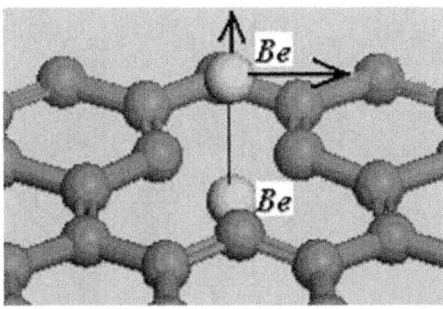

Figure 22: Scheme of the elastic modulus calculation for the Be dumbbell at vacancy.

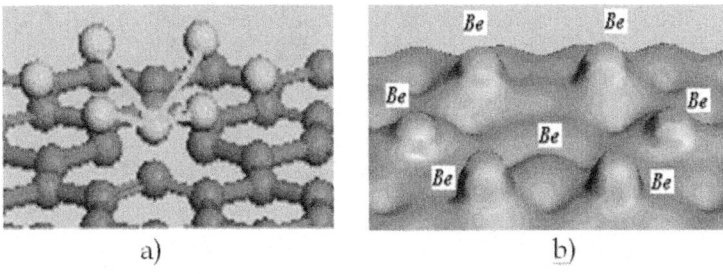

a)   b)

Figure 23: a) A configuration of an initially flat Be-cluster over vacancy after optimization. The binding energy of the cluster with graphene was obtained as large as 7.2 eV; b) the electron charge distribution in the area of the metal cluster – vacancy with a density of charge 0.2 el / $Å^3$.

AL – GRAPHENE COMPOSITION

A stable configuration with a minimum of total energy was provided by Al atom placed in the plate of graphene sheet within the vacancy. One can see in Figure 24 that the vacancy zone essentially and symmetrically increased with all three lengths of bonds equal 1.7 Å.

Figure 25 presents configuration of Al – dumbbell placed at the vacancy. The equilibrium distance between Al atoms equals 2.5 Å, the binding energy is as large as 2.9 eV.

Very interesting and important result for technological applications by production of composite materials based on Al or Al –alloys matrix with graphene filler shows computational model in Figure 27. One can see, that Al atoms can be chemically attached to a bridge-like defect in vacancy's zones. One also can see a significant deformation of graphene sheets around the defect zone. The binding energy of the interstitial carbon atom i at this configuration was equal to -9.3 eV.

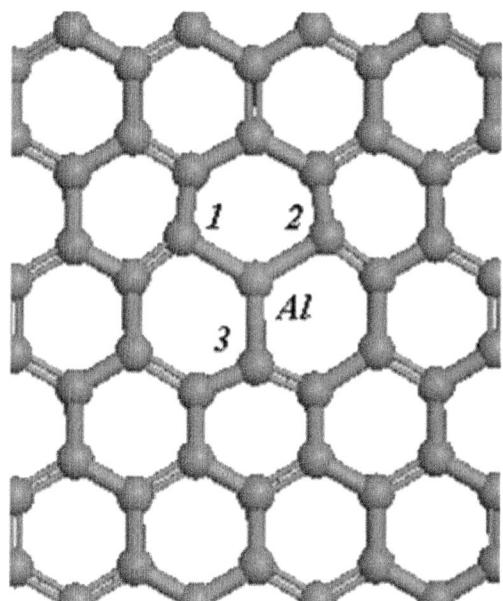

Figure 24: Configuration of a single Al-atom bonded with a vacancy.

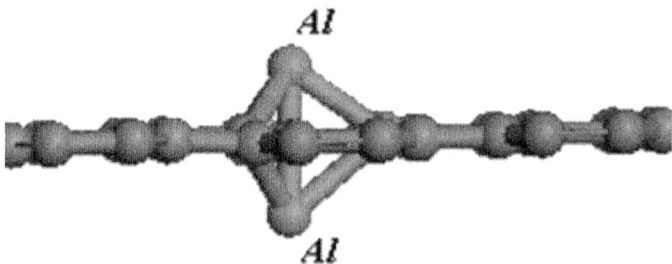

Figure 25: Configuration of Al – dumbbell bonded with a vacancy.

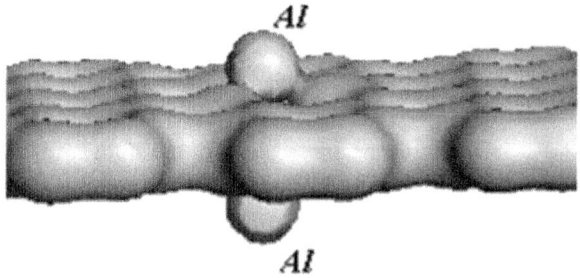

Figure 26: The distribution of the electron charge (the density equals 1.4 el / A³) for the Al – dumbbell bonded with a vacancy in graphene.

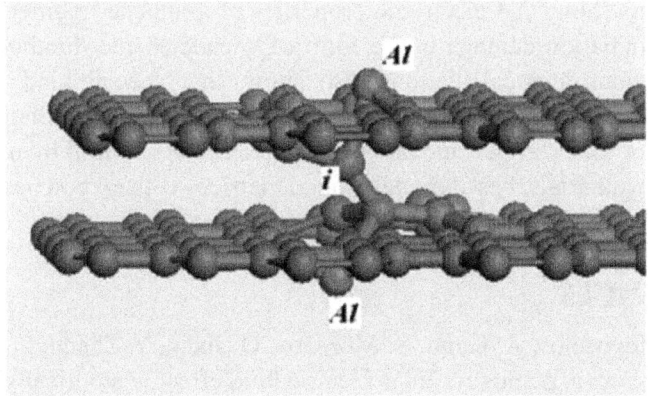

Figure 27: Complex defect: a bridge-like defect in two- layer graphene with Al atoms attached on both graphene sheets.

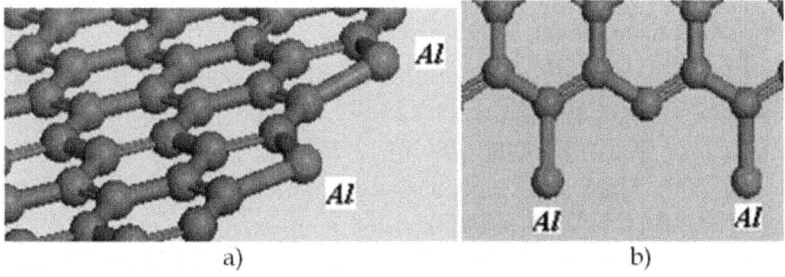

a)   b)

Figure 28: Edge bonds of graphene fragments with Al atoms caught with binding energy E_b as large as 4.2 eV (a) and 3.5 eV (b).

Figure 28 illustrates one more possible way of creation additional chemical bonds between Al atoms and graphene fragments. Free end bonds can serve as additional sites of linking Al atoms providing better adhesion between metal matrix and graphene filler with binding energy as large as 4.2 eV in the case of "arm-chair" edge and 3.5 eV for "zig-zag" edge.

CONCLUSION

Some stable radiation defect configurations, involving single adsorbed carbon atom, 3D C-dumbbell defect and a vacancy-like defect in graphene sheet, associated with a dumbbell of adsorbed atoms as well as complex interior defects : bridge-like radiation defects, which can originate under fast electron or ions irradiation and attach metal atoms, were simulated. The binding energy, structure characteristics and some electron characteristics of defects have been determined by using molecular dynamics, MO LCAO and DFT. Our calculations show, that electronic properties of graphene fragments are rather stable to radiation damage in the form of vacancies and dumbbells. Results of simulations and calculations also show, that special kind of radiation defects, namely bridge-like radiation defects produced by irradiation with fast electrons or ions can become an effective technological tool by production of composite materials, based on light metal matrices with carbon nanostructures, in particular, graphene's fragments, as reinforcement elements.

REFERENCES

1. K. Novoselov, A. Geim, S. Morozov, D. Jiang, Y. Zhang, S. Dubonos, I. Grigorieva, A. Firsov, 2004 Electric field effect in atomically thin carbon films. Science, 306 666 669 , 0036-8075

2. D. Elias, R. Nair, T. Mohiuddin, S. Morozov, P. Blake, M. Halsall, A. Ferrari, D. Boukhvalov, M. Katsnelson, A. Geim, K. Novoselov, 2009 Control of Graphene's Properties by Reversible Hydrogenation: Evidence for Graphane. Science, 323 610 613 .

3. J. Sofo, A. Chaudhari, G. Barber, 2007 Graphane: a two-dimensional hydrocarbon. Phys.Rev.,B75, 153401 153404 , 0974-0546

4. Zh. Luo, T. Yu, K. Kim, Zh. Ni, Y. You, S. Lim, Z. Shen, S. Wang, J. Lin, 2009 Thickness-Dependent Reversible Hydrogenation of Graphene Layers. ACS NANO, 3 7 1781 1788 .

5. D. Teweldebrhan, A. Balandin, 2009 Modification of graphene properties due to electron- beam irradiation. Appl.Phys.Lett., 94 013101

6. A. Ilyin, E. Daineko, G. Beall, 2009 Computer simulation and study of radiation defects in graphene. Physica E, 42 No 67 69 , 1386-9477

7. A. Ilyin, N.. Guseinov, A. Nikitin, I. Tsyganov, 2010 Characterization of thin graphite layers and graphene with energy dispersive X-ray analysis. Physica E, 42 8 2078 2080 .

8. Ilyin,A. 2010 Simulation of end-bridge-like radiation defects in carbon multi-wall nanotubes. Book of Abstracts of 10th International Conference on Computer Simulations of Radiation Effects in Solids, 123 ISBN, Poland, Krakov, July 19-23, 2010.

9. Ilyin,A &.Beall,G. 2010 Computer simulation and study of bridge-like radiation defects in the carbon nano-structures in composite materials. Proceedings of NanoTech Conference, 312 315 , 978-1-43983-401-5 Annaheim, California, USA, June, 21-25, 2010.

10. A. Ilyin, G. Beall, I. Tsyganov, 2010 Simulation and Study of Bridge-Like Radiation Defects in the Carbon Nano-Structures. Journal of

Computational and Theoretical Nanoscience, 7 10 Oct.,2010),2004 2007
, 1546-1955

11. A. Ilyin, N. Guseinov, I. Tsyganov, R. Nemkaeva, 2011 Computer
Simulation and Experimental Study of Graphan-Like Structures Formed
by Electrolytic Hydrogenation. Physica E, 43 1262 1265 ,1386-9477

Chapter 8

COMPUTER SIMULATION OF MOLECULAR DYNAMICS:METHODOLOGY, APPLICATIONS, AND PERSPECTIVES IN CHEMISTRY

Wilfred E van Gunsteren[1] and Herman J. C. Berendsen[2]

[1]Department of Physical Chemistry University of Groningen Nijenborgh 16, NL-9747 AG Groningen (The Netherlands)

[2]Department of Physical Chemistry University of Groningen Nijenborgh 16, NL-9747 AG Groningen (The Netherlands

ABSTRACT

During recent decades it has become feasible to simulate the dynamics of molecular systems on a computer. The method of molecular dynamics (MD) solves Newton's equations of motion for a molecular system, which results in trajectories for all atoms in the system. From these atomic trajectories a variety of properties can be calculated. The aim of computer simulations of molecular systems is to compute macroscopic behavior from microscopic interactions. The main contributions a microscopic consideration can offer are (1) the understanding and (2) interpretation of experimental results, (3) semiquantitative estimates of experimental results, and (4) the capability to interpolate or extrapolate experimental data into regions that are only difficultly accessible in the laboratory. One of the two basic problems in the field of molecular modeling and simulation is how to efficiently search the vast configuration space which is spanned by all possible molecular conformations for the global low (free) energy regions which will be populated by a molecular system in thermal equilibrium. The other basic problem is the derivation of a sufficiently accurate interaction energy function or force field for the molecular system of interest. An important part of the art of computer simulation is to choose the unavoidable assumptions, approximations and simplifications of the molecular model and computational procedure such that their contributions to the overall inaccuracy are of comparable size, without affecting significantly the property of interest. Methodology and some practical applications of computer simulation in the field of (bio)chemistry will be reviewed.

INTRODUCTION

Computational chemistry is a branch of chemistry that enjoys a growing interest from experimental chemists. In this discipline chemical problems are resolved by computational methods. A model of the real world is constructed, both measurable and unmeasurable properties are computed, and the former are compared with experimentally determined properties. This comparison validates or invalidates the model that is used. In the former case the model may be used to study relationships between model parameters and assumptions or to predict unknown or unmeasurable quantities. Since chemistry concerns the study of properties of substances or molecular systems in terms of atoms, the basic challenge facing computational chemistry is to describe or even predict 1. the structure and stability of a molecular system, 2. the (free) energy of different states of a molecular system, 3. reaction processes within molecular systems in terms of interactions at the atomic level. These three basic challenges are listed according to increasing difficulty. The first challenge concerns prediction of which state of a system has the lowest energy. The second challenge goes further; it involves prediction of the relative (free) energy of different states. The third challenge involves prediction of the dynamic process of change of states.

Chemical systems are generally too inhomogeneous and complex to be treated by analytical theoretical methods. This is illustrated in Figure 1. The treatment of molecular systems in the gas phase by quantum mechanical methods is straightforward; if a classical statistical mechanical approximation is permitted the problem becomes even trivial. This is due to the possibility of reducing the many-particle problem to a few-particle one based on the low density of a system in the gas phase. In the crystalline solid state, treatment by quantum mechanical or classical mechanical methods is made possible by a reduction of the many-particle problem to a few-(quasi)particle problem based on symmetry properties of the solid state. Between these two extremes, that is, for liquids, macromolecules, solutions, amorphous solids, etc., one is faced with an essentially many-particle system. No simple reduction to a few degrees of freedom is possible, and a full treatment of many degrees of freedom is required in order to adequately describe the properties of molecular systems in the fluid-like state. This state of affairs has two direct consequences when treating fluid-like systems.

1. One has to resort to numerical simulation of the behavior of the molecular system on a computer, which

2. produces a statistical ensemble of configurations representing the state of the system.

If one is only interested in static equilibrium properties, it suffices to generate an ensemble of equilibrium states, which may lack any temporal correlations. To obtain dynamic and non-equilibrium properties dynamic simulation methods that produce trajectories in phase space are to be used. The connection between the microscopic behavior and macroscopic properties of the molecular system is governed by the laws of statistical mechanics. Figure I also shows the broad applicability of computer simulation methods in chemistry. For any fluid-like, essentially many-particle system, it is the method of choice.

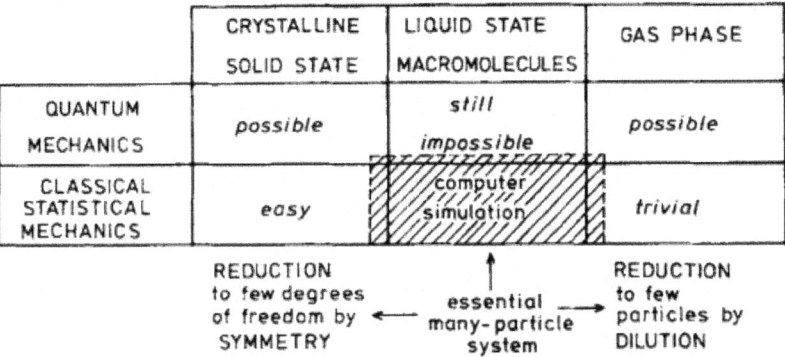

Figure 1: Classification of molecular systems. Systems in the shaded area are amenable to treatment by computer simulation.

The expanding role of computational methods in chemistry has been fueled by the steady and rapid increase in computing power over the last 40 years, as is illustrated in Figure 2. The ratio of performance to price has increased an order of magnitude every 5-7 years, and there is no sign of any weakening in this trend. The introduction of massive parallelism in computer architecture will easily maintain the present growth rate. This means that more complex molecular systems may be simulated over longer periods of time, or that it will be possible to handle more complex interaction functions in the decades to come. The present article is concerned with computer circulation of molecular systems. In Section 2 the two basic problems are formulated, a brief history of dynamic computer simulation is presented, the reliability of current simulations is discussed, and the usefulness of simulation studies is considered. Section 3 deals with simulation methodology: choice of computational model and atomic interaction function (Section 3.1), techniques to search configuration space for low energy configurations (Section 3.2), boundary conditions (Section 3.3), types of dynamical simulation methods (Sec tion 3.4), and algorithms for integration of the equations of motion (Section 3.9, and finally equilibration and analysis of molecular systems (Section 3.6) In Section 4, a number of applications of computer simulation in chemistry are dis

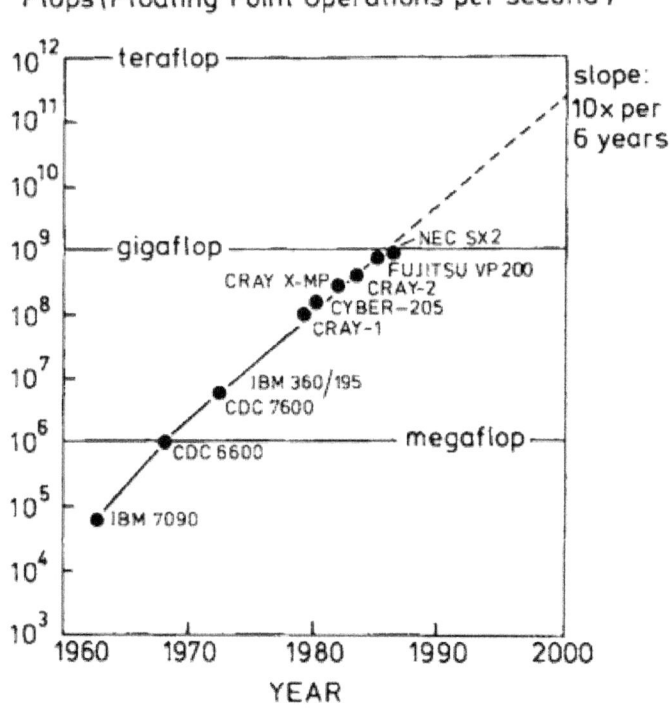

Figure 2: Development of computing power of the most powerful computers.

Cussed and examples are given. Finally, future developments are considered in Section 5. For other relatively recent monographs on computer simulation the reader is referred to the literature (see Refs. 1-91).

COMPUTER SIMULATION OF MOLECULAR SYSTEMS

Two Basic Problems

Two basic problems are encountered in the computer simulation of fluid-like molecular systems:

1. the size of the configurational space that is accessible to the molecular system, and

2. the accuracy of the molecular model or atomic interaction function or force field that is used to model the molecular system.

Table 1: Models at different levels of approximation

Model	Degrees of freedom		Example of		
	left	removed	predictable property	force field	
– quantum mechanical	nuclei, electrons	nucleons	reactions	Coulomb	increasing: simplicity, speed of computation, search power, time scale
– all atoms, polarizable	atoms dipoles	electrons	binding charged ligand	ionic models [10]	
– all atoms	solute + solvent atoms	dipoles	hydration	OPLS [11] GROMOS [12]	
– all solute atoms	solute atoms	solvent	gas phase conformation	MM2 [13]	decreasing: complexity, accuracy of atomic properties
– groups of atoms as balls	atom groups	individual atoms	folding topology of macromolecule	LW [14]	

Size of the Configurational Space

The simulation of molecular systems at non-zero temperatures requires the generation of a statistically representative set of configurations, a so-called ensemble. The properties of a system are defined as ensemble averages or integrals over the configuration space (or more generally phase space). For a many-particle system the averaging or integration will involve many degrees of freedom, and as a result can only be carried out over part of the configuration space. The smaller the configuration space, the better the ensemble average or integral can be approximated. When choosing a model from which a specific property is to be computed, one would like to explicitly include only those degrees of freedom on which the required property is dependent. In Table 1 a hierarchy of models is shown, in which specific types of degrees of freedom are successively removed. Examples are given of properties that can be computed at the different levels of approximation. If one is interested in chemical reactions, a quantum mechanical treatment of electronic degrees of freedom with the nuclear coordinates as parameters is mandatory. The intranuclear degrees of freedom (nucleon motion) are left out of the model. The computing power required for a quantum mechanical treatment scales at least with the third power of the number of electrons N, that are considered. Such a treatment is only possible for a limited number of degrees of freedom. At a next level of approximation the electronic degrees of freedom are removed from the model and approximated by (point) polarizabilities. Such a model, allowing for motions of polarizable atoms, could, e.g., be used to compute the binding properties of polar or charged ligands which will polarize the receptor. At this level of modeling the computing effort scales with the square of the number of atoms N_a. This means that many more atoms can be considered than at the quantum level. Removal of the polarizability of the atoms yields a next level of approximation, in which only atomic positional degrees of freedom are considered and the mean polarization is included in the effective interatomic interaction function. Such a model allows for the study of solvation properties of molecules. When studying solutions, a next level of approximation is reached by omitting the solvent degrees of freedom from the model and simultaneously adapting the interaction function for the solute such that it includes the mean solvent effect. When studying protein folding, the configuration space spanned by all increasing: simplicity, speed of computation, search power, time scale decreasing: complexity, accuracy of atomic properties protein atoms is still far too large to be searched for low energy conformations. In this case a further reduction in it can be obtained by representing whole groups of atoms, for example an amino acid residue, as one or two balls (3 or 6 degrees of freedom).

From this discussion it is clear that the level of approximation of the model that will be used in the simulation will depend on the specific property one is interested in. The various force fields that are available correspond to different levels of approximation, as is illustrated in the right half of Table 1. There is a hierarchy of force fields. Once the level of approximation has been chosen, that is, the types of degrees of freedom in the model, one must decide how many degrees of freedom (electrons, atoms, etc.) are to be taken into account. How small a system can be chosen without seriously affecting a proper representation of the property of interest? The smaller the size of the system the better its degrees of freedom can be sampled, or in dynamic terms, the longer the time scale over which it can be simulated. We may summarize the first basic problem in the computer simulation of molecular systems as follows: the level of approximation of the model should be chosen such that those degrees of freedom that are essential to a proper evaluation of the quantity or property of interest can be sufficiently sampled. In practice any choice involves a compromise between type and number of degrees of freedom and extent of the simulation on the one hand, and the available computing power on the other.

Accuracy of Molecular Model and Force Field

When the degrees of freedom are infinitely dense or long sampled the accuracy by which various quantities are predicted by a simulation will depend solely on the quality of the assumptions and approximations of the molecular model and interatomic force field. At the level (Table 1) of quantum mechanical modeling the basic assumption is the validity of the Born-Oppenheimer approximation separating electronic and nuclear motion. The interaction between point atoms and electrons is described by Coulomb's law and the Pauli Exclusion Principle. When excluding chemical reactions, low temperatures or details of hydrogen atom motion, it is relatively safe to assume that the system is governed by the laws of classical mechanics. The atomic interaction function is called an effective interaction since the average effect of the omitted (electronic) degrees of freedom has been incorporated in the interaction between the (atomic) degrees of freedom explicitly present in the model. To each level of approximation in Table 1 there corresponds a type of effective interaction or force field. For example, a pair potential with an enhanced dipole moment (enhanced atomic charges) may be used as an effective potential that mimics the average effect of polarizability.

In view of the different levels of approximation of molecular models it is not surprising that the literature contains a great variety of force fields. They can be classified along different lines:

different lines: type of compound that is to be mimicked, e.g. carbohydrates, sugars, polypeptides, polynucleotides ;

- type of environment of the compound of interest; e.g. gas phase, aqueous or nonpolar solution;
- range of temperatures covered by the effective interaction;
- type of interaction terms in the force field; e.g. bond stretching, bond angle bending, torsional terms, two-
- three- or many-body nonbonded terms;
- functional form of the interaction terms; e.g. exponential or 12th power repulsive nonbonded interaction;
- type of parameter fitting, that is, to which quantities are parameters fitted, and are experimentally or ab-initio theoretically obtained values used as target values.

We note that the choice of a particular force field should depend on the system properties one is interested in. Some applications require more refined force fields than others. Moreover, there should be a balance between the level of accuracy or refinement of different parts of a molecular model. Otherwise the computing effort put into a very detailed and accurate part of the calculation may easily be wasted due to the distorting effect of the cruder parts of the model.

Assumptions, Approximations and Limitations

When using a particular model to predict the properties of a molecular system, one should be aware of the assumptions, simplifications, approximations and limitations that are irnplicit in the model. Below, we list the four most important approximations and limitations of classical computer simulation techniques which should be kept in mind when using them.

Classical Mechanics of Point Masses

A molecular system is described as a system of point masses moving in an effective potential field, which is generally a conservative field, i.e. it only depends on the instantaneous coordinates of the point masses. The motion of the point masses is with a sufficient degree of accuracy governed by the laws of classical mechanics. These assumptions imply the following restrictions to modeling:

*   low-temperature (0- 10 K) molecular motion is not ade-
*   detailed motion of light atoms such as hydrogen atoms is
*   Description of chemical reactions lies outside the scope of quately described; not correctly described even at room temperature; classical simulation methods.

For a short discussion of the inclusion of quantum corrections to a classical treatment or of the quantum (dynamical) simulation techniques that are presently under development, we refer to Section 5.1.

System Size or Number of Degrees of Freedom to be Included

Only a rather limited number of atoms can be simulated on a computer. Simulations of liquids typically involve 10'- 103 atoms, simulations of solutions or crystals of macromolecules about 103-2 x lo4 atoms. Generally, the system size is kept as small as possible in order to allow for a sufficient sampling of the degrees of freedom that are simulated. This means that those degrees of freedom that are not essential to the property one is interested in should be removed from the system. The dependence of the property of interest on the size of the system may give a clue to the minimum number of degrees of freedom required for an adequate simulation of it. The larger the spatial correlation length of the property of interest, the more atoms are to be included in the simulation.

Sufficiency of Sampling of Configuration Space or Time Scale of Processes

Computer simulation generates an ensemble of configurations of the system. Whether the generated set of configurations is representative for the state of the system depends on the extent to which the important (generally low energy) parts of the configuration space (or, more generally formulated, phase space) have been sampled. This depends in turn on the sampling algorithm that is applied. This should be able to overcome the multitude of barriers of the multidimensional energy surface of the system. In dynamic simulations, the time scale of the process that is mimicked is limited. Presently, molecular simulations cover time periods of 100- 1000 picoseconds. In the case of activated processes longer time scales can be reached by using special tricks. Essentially slow processes, like the folding of a protein, are still well out of reach of computer simulation. We note that the observation that a property is independent of the length of a simulation is a necessary but not a sufficient condition for adequate sampling. The system may just reside for a period longer than the simulation time in a certain area without being effected by a nearby region of much lower energy from which it is separated by a large energy barrier.

Accuracy of the Molecular Model and Force Field

As is illustrated in Table 1, there exists a variety of molecular models and force fields, differing in the accuracy by which different physical quantities are modeled. The choice of a particular force field will depend on the property and level of accuracy one is interested in. When studying a molecular system by computer simulation three factors should be considered (cf. Fig. 3).

1. The properties of the molecular system one is interested in should be listed and the configuration space (or time scale) to be searched for relevant configurations should be estimated.
2. The required accuracy of the properties should be specified.
3. The available computer time should be estimated.

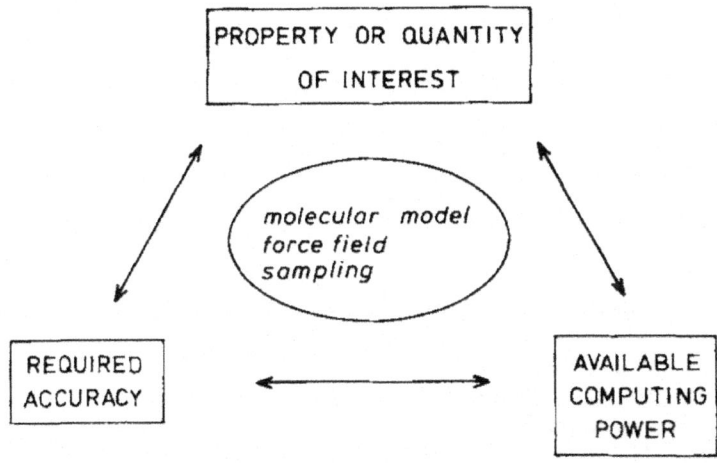

Figure 3: Choice of molecular model, force field and sample size depends on 1) the property one is interested in (space to be searched), 2) required accuracy of the prediction, 3) the available computing power to generate the ensemble.

Given these three specifications the molecular model and force field should be chosen. As is indicated in Table 1 there is a trade off between accuracy of the force field on the one hand and the searching power or time scale that can be attained on the other. We note that in many practical cases one should abandon the idea of trying to simulate the system of interest, since the available computing power will not allow for a sufficiently accurate simulation.

History of Dynamic Computer Simulation

The growth in the number of applications of computer simulation methods in chemistry and physics is directly caused by the rapid increase in computing power over the last four decades (Fig. 2). Table 2 shows the development of the application of molecular dynamics simulation in chemistry. Alder and Wainwright pioneered the method using 2-dimensional hard disks. Rahman, who can really be considered the father of the field, simulated liquid argon in 1964, liquid water in 1971, and superionic conductors in 1978.

271 He also made many contributions to the methodology, and stimulat ed the dissemination of the technique in the scientific community. In the seventies, the transition from atomic to molecular liquids was made: rigid molecules like water (1971), flexible alkanes (1975), and a small protein, trypsin inhibitor (1977). The application to molten salts (1971) required the development of methods to handle long-range Coulomb interactions. The eighties brought simulations of biomolecules of increasing size in aqueous solution. For a more thorough review of the past and present of dynamic computer simulation in chemistry and physics we refer to the monographs of McCammon and Harveyr5I and Allen and Tildesley.

Table 2: History of computer simulation of molecular dynamics

Year	System	Length of simulation [s]	Required cpu time on super-computer [h]
1957	hard two-dimensional disks [17]		
1964	monatomic liquid [18]	10^{-11}	0.05
1971	molecular liquid [19]	5×10^{-12}	1
1971	molten salt [20]	10^{-11}	1
1975	simple small polymer [21]	10^{-11}	1
1977	protein in vacuo [22]	2×10^{-11}	4
1982	simple membrane [23]	2×10^{-10}	4
1983	protein in aqueous crystal [24]	2×10^{-11}	30
1986	DNA in aqueous solution [25]	10^{-10}	60
1989	protein-DNA complex in solution [26]	10^{-10}	300
	large polymers	10^{-8}	10^3
	reactions	10^{-4}	10^7
	macromolecular interactions	10^{-3}	10^8
	protein folding	10^{-1}	10^9

From Table 2 it is also clear that a supercomputer is still many orders of magnitude slower than nature: a state of the art simulation is about 1015 times slower than nature. Many interesting systems and processes still fall far outside the reach of computer simulation. Yet, with a growth rate of a factor 10 per 5-7 years for (super) computing power the speed of simulation will have caught up with that of nature in about 100 years.

Accuracy and Reliability of Computer Simulation

The reliability of predictions made on the basis of computer simulation will basically depend on two factors: 1) whether the molecular model and force field are sufficiently accurate, and 2) whether the configuration space accessible to the molecular system has been sufficiently thoroughly searched for low-energy configurations. Although the accuracy of a prediction may be estimated by considering the approximations and simplifications of the model and computational procedure, the final test lies in a comparison of theoretically predicted and experimentally measured properties. In order to provide a firm basis for the application of computer simulation methods the results should be compared with experimental data whenever possible. In this context we would like to stress that good agreement between calculated and experimental data does not necessarily mean that the theoretical model underlying the calculation is correct. Good agreement may be due to any of the following reasons:

1. The model is correct, that is, any other assumption used to derive the model, or any other choice of parameter values would give bad agreement with experiment.

2. The property that is compared is insensitive to the assumptions or parameter values of the model, that is, whatever parameter values are used in the model calculation, the agreement with experiment will be good.

3. Compensation of errors occurs, either by chance or by fitting of the model parameters to the desired properties.

A number of examples of the last case, viz. good agreement for the wrong reason, are given in Ref. Here, we would only like to remind the reader that it is relatively easy, when modeling high-dimensional systems with many parameters, to choose or fit parameters such that good agreement is obtained for a limited number of observable quantities. On the other hand, disagreement between simulation and experiment may also have different causes: 1) the simulation is not correct, the experiment is, 2) the simulation is correct, the experiment is not.

With these cautionary remarks in mind we would like to turn to an evaluation of the accuracy of current simulation studies by comparing simulated with experimental quantities. Table 3 contains a scheme of the atomic quantities of molecular systems for which comparison between simulated and experimental values is feasible. Four phases are distinguished. The most interesting phase from a chemical point of view is that of a molecule in solution. For this phase only few accurately measured atomic properties are available for comparison, leaving thermodynamic system properties like the free energy of solvation as a test ground for the simulation. Considerably more atomic data are available from crystallographic diffraction experiments : atomic positions and mobility, although the accuracy of the latter is much lower than that of the former, due to the simplifying approximations (harmonic isotropic motion) used in the crystallographic refinement process

Table 3: Possible comparison of simulated properties with experimental ones for complex molecules

Atomic properties	Experimental method	Phase			
		gas	solution	membrane	crystal
Structure					
– positions	{X-ray diffraction				×
	{neutron diffraction				×
– distance	NMR		×		
– orientations	NMR			×	
Mobility					
– B-factors	{X-ray diffraction				×
	{neutron diffraction				×
– occupancy factors	{X-ray diffraction				×
	{neutron diffraction				×
Dynamic properties					
– vibrational frequencies	infrared spectroscopy	×			
– relaxation rates	various NMR optical techniques		×		
– diffusion	NO spin label		×	×	
Thermodynamic properties					
– density			×		×
– free energy			×		
– viscosity, conductance			×		

In the following subsections examples will be given of a comparison of various atomic and system properties for different compounds. The examples are taken from our own work and so are all based on the GROMOS force field. ['2' They only serve as an indication of the degree of accuracy that can be obtained for different properties using current modeling methods.

Atomic Properties: Positions and Mobilities in Crystals

From a simulation the average molecular structure can be easily calculated and compared with experiment. Table 4 contains the deviation between simulated and measured atomic positions for a number of molecular crystals. The structural deviation depends on the size of the molecule and ranges from 0.2 A to 1.2A. The numbers are an average over part or over all atoms in the molecules. This means that parts of the molecule will deviate more and other parts will deviate less from the experimental structure. This is illustrated in Figure 4, which shows the deviation for the backbone C, atoms as a function of residue number averaged over all four BPTI (bovine pancreatic trypsin inhibitor) molecules in the crystal unit cell. Most of the atoms deviate less than 1 A. Figure 5 shows the root mean square (rms) atomic positional fluctuations calculated from the MD simulation and from the crystallographic B-factors. The largest discrepancy is still rather small, about 0.3 A.

Table 4: Comparison of MD time-averaged structures with experimental X-ray or neutron diffraction structures.

Molecule	Size (number of glucose units or amino acid residues)	Root mean square (RMS) difference in atomic positions [Å]	
		C_a or C_3 atoms	all atoms excl. H atoms
cyclodextrin (α/β) [29]	6	0.13/0.35	0.25/0.51
cyclosporin A [30]	11	0.3	0.6
trypsin inhibitor [31]	58	1.0	1.5
subtilisin [32]	275	1.0	1.2

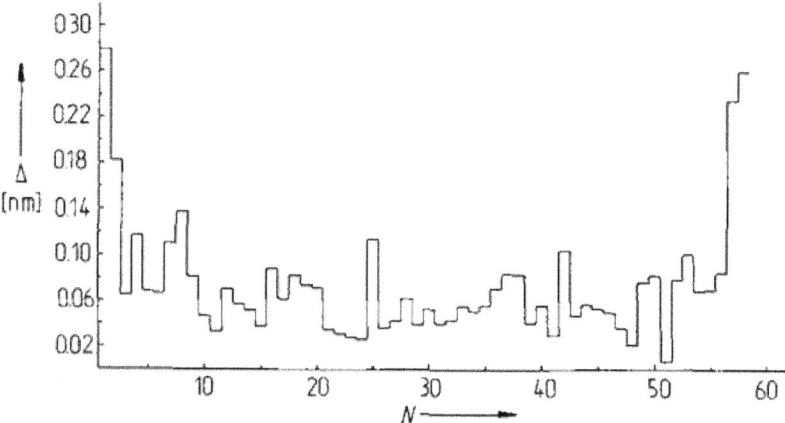

Figure 4: Root mean square difference between the trypsin inhibitor simulated time- and molecule-averaged C. atomic positions and the X-ray positions according to 31.. A = root mean square deviation, N = aminoacid residue number.

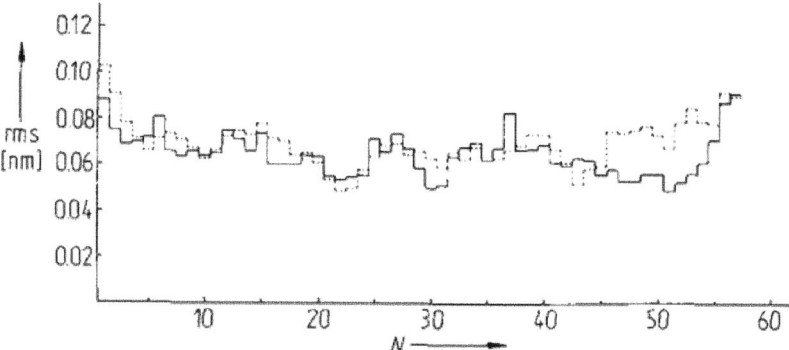

Figure 5: Root mean square positional fluctuations for the trypsin inhibitor atoms averaged over the backbone atoms of each residue according to 1311. Solid line: simulated results. Broken line: fluctuations derived from a set of X-ray temperature factors. rms = root mean square fluctuation, N = amino acid residue number.

A more stringent test of the simulation is to check whether partially occupied atomic sites are reproduced. The neutron diffraction work on p-cyclodextrin shows that 16 hydrogen atoms occupy two alternative sites. In the crystal simulation a non-zero occupancy is observed for 84 % of the hydrogen sites, and for 62 YO of the hydrogen atoms the relative occupancy of the two alternative sites is qualitatively reproduced

The molecular structure can also be described in terms of hydrogen bonds. Generally the experimentally observed hydrogen bonds are also observed in the simulation. Again cyclodextrin crystals form a sensitive test case for the simulations, since peculiar geometries such as three-center hydrogen bonds, and dynamic processes such as flip-flop hydrogen bonds have been experimentally observed. Almost all experimentally observed three-center hydrogen bonds in crystals of CL- and P-cyclodextrin are reproduced in MD simulation, even as far as the detailed asymmetric geometry is con- ~erned.'~~. In a so-called flip-flop hydrogen bond the directionality is inverted dynamically. In a MD simulation of B-cyclodextrin, 16 of the 18 experimentally detected flip-flop bonds are reproduced.

Atomic Properties: Distances in Solution

Structural data of solutions can be obtained by twodimensional nuclear magnetic resonance spectroscopy (2DNMR) with exploitation of the nuclear Overhauser (NOE) experiments.

351 The data come in the form of a set of upper bounds or constraints to specific proton-proton distances. This affords the possibility of comparing these proton-proton distances as predicted in a simulation with the experimentally measured NOE bounds. In the literature,

251 such a comparison is given for a set of 174 NOE's in an eight basepair DNA fragment in aqueous solution: 80% of the NOE distances are satisfied by the simulation within experimental error, the mean deviation is 0.22 A, and the maximum deviation amounts to 2.9 A.

Atomic Properties: Orientation of Molecular Fragments in Membranes

The degree of order in a membrane or lipid bilayer can be measured selectively along the aliphatic chain by deuterium NMR spectroscopy. In Figure 6 both experimental and MD order parameters are displayed for all CH, units in a sodium ion/water/decanoate/decanol bilayer system as a function of the position of the carbon atom in the aliphatic chain. The experimentally observed plateau in the order parameters is well reproduced.

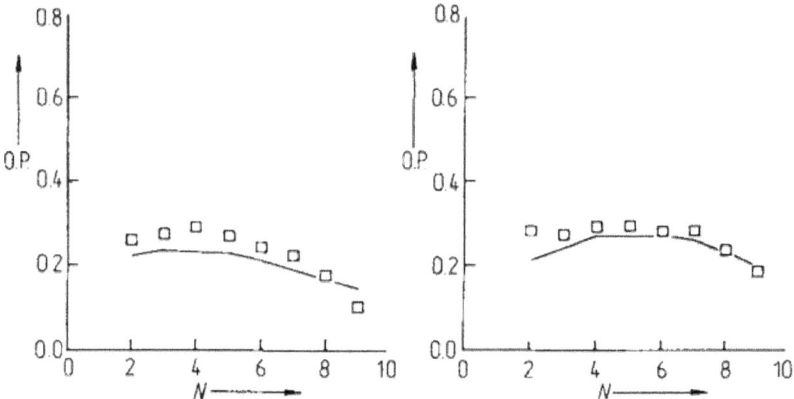

Figure 6: Carbon-deuterium order parameters as a function of carbon atom number along the aliphatic chain of decanol (upper panel) and decanoate (lower panel)

The head group has number zero. Solid line: simulated result. Squares: experimental values. O.P. = order parameter, N = carbon atom number.

Atomic Properties: Diffusion in Solutions and in Membranes

The dynamic properties of simulations of molecular systems are hard to test by comparison with experiment. An exception is the diffusion constant. Its calculation and comparison with experiment is a standard test for models of liquid water. The diffusion constant of the simple three-point charge (SPC) water model is 4.3 x 10^{-5}cm2 s-' compared with an experimental value of 2.7×10^{-5}cm^{2}s^{-1} at 305 K.

381 Inclusion of a correction term for self-polarization in the SPC model led to a reparametrization, to the SPC/E (extended simple point charge) model, which yields a considerably smaller diffusion constant of 2.5 x lo^{-5} cm2 s-1.

391 In the bilayered sodium ion/water/decanoate/decanol system mentioned in the previous section the simulated diffusion constants are: 2.7 x10^{-6}cm^2 s^{-1} (decanoate) and 5.2 x10^{-6}cm^2 s^{-1} (decanol), which values are to be compared with the value of 1.5 x10^{-6}cm^2 s^{-1} measured with nitroxide spin labels.

System Properties: Thermodynamic Quantities

Instead of atomic properties, thermodynamic system properties like the density or free energy of solvation can be calculated from a simulation and compared to experimental values. For a crystal of a cytidine derivative X-ray diffraction data and MD simulation data were compared at two different temperatures.

401 Upon raising the temperature from 11 3 K to 289 K the volume of the unit cell increased by 2.5 % in the simulation at constant pressure, which value is to be compared with an experimentally determined increase of 3.7%. At 289 K the simulated density was only 1.3% too large. A slightly larger discrepancy was found in a MD simulation of crystalline cyclosporin A: the simulated density of 1.080 g cm^{-3} was 3.6% larger than the experimental value of 1.042 g cm^{-3}.

1. Comparable deviations have been found for other systems: the SPC/E model yields 0.998 g cm^{-3} at 306K, compared with an experimental value of 0.995 g cm^{-3} at 305 K

2. Finally, we quote an example of a free energy of hydration. For the solvation of methanol in water the standard GROMOS parameters yield a free enthalpy of 7 kJ mol^{-1} compared with an experimental value of 5 kJ mol- *.

3. However, when charged moieties are involved the accuracy by which free energy of solvation can be calculated is no better than about 10-20 kJ mol^{-1}.

Is Computer Simulation Useful?

The prediction of properties by computer simulation of complex molecular systems is certainly not accurate enough to justify abandoning the measurement of properties. If a measurement is not too difficult it is always to be preferred over a prediction by simulation. The utility of computer simulation studies does not lie in the (still remote) possibility of replacing experimental measurement, but rather in its ability to complement experiments. Quantities that are inaccessible to experiment can be monitored in computer simulations.

We consider the computer simulation of complex molecular systems to be useful for the following reasons.

1. It provides an understanding of the relation between microscopic properties and macroscopic behavior. In a computer a microscopic molecular model and force field can be changed at will and the consequences for the macroscopic behavior of the molecular system can be evaluated.

2.During the last few years computer simulation has become a standard tool in the determination of spatial molecular structure on the basis of X-ray diffraction or 2D-NMR data.

3. Under favorable conditions computer simulation can be used to obtain quantitative estimates of quantities like binding constants of ligands to

receptors. This is especially useful where the creation of the ligand or the measurement of its binding constant is costly or time-consuming.

4. Finally we mention the possibility of carrying out simulation under extreme (unobservable) conditions of temperature and pressure.

METHODOLOGY

Here we briefly describe the methodology of classical computer simulation as it is applied to complex molecular systems.

Choice of Molecular Model and Force Field

Current Force Fields for Molecular Systems

A typical molecular force field or effective potential for a system of N atoms with masses m, (i = 1,2,. . .,N) and cartesian position vectors r; has the form (1).

$$V(r_1, r_2, \ldots, r_N) = \sum_{\text{bonds}} \tfrac{1}{2} K_b [b - b_0]^2 + \sum_{\text{angles}} \tfrac{1}{2} K_\theta [\theta - \theta_0]^2$$

$$+ \sum_{\substack{\text{improper} \\ \text{dihedrals}}} \tfrac{1}{2} K_\xi [\xi - \xi_0]^2$$

$$+ \sum_{\text{dihedrals}} K_\varphi [1 + \cos(n\varphi - \delta)]$$

$$+ \sum_{\text{pairs}(i,j)} [C_{12}(i,j)/r_{ij}^{12} - C_6(i,j)/r_{ij}^6$$

$$+ q_i q_j / (4\pi\varepsilon_0\varepsilon_r r_{ij})] \tag{1}$$

The first term represents the covalent bond stretching interaction along bond b. It is a harmonic potential in which the minimum energy bond length b, and the force constant Kb vary with the particular type of bond. The second term describes the bond angle bending (three-body) interaction in similar form. Two forms are used for the (four-body) dihedral angle interactions: a harmonic term for dihedral angles 5 that are not allowed to make transitions, e.g. dihedral angles within aromatic rings, and a sinusoidal term for the other dihedral angles c_p, which may make 360 degree turns. The last term is a sum over all pairs of atoms and represents the effective nonbonded interaction, composed of the van der Waals and the Coulomb interaction between atoms i and j with charges qi and q_j at a distance r_{ij}.

There exists a large number of variants of expression (I)." - 13.42- 531

Some force fields contain mixed terms like $K_{b\theta}[b - b_0][\theta - \theta_0]$, which directly couple bond-length and bond-angle vibrations.Others use more complex dihedral interactions terms.The choice of the relative dielectric constant ε_r is also a matter of dispute. Values ranging from $\varepsilon_r = 1$ $\varepsilon_r = 8$ have been used, while others take c, proportional to the distance r_{ij}, Sometimes the Coulomb term is completely ignored. Although hydrogen bonding can be appropriately modeled using expression (1),in some force fields special hydrogen bonding potential terms are used to ensure proper hydrogen bonding. An other way to refine expression (1) is to allow for non-atomic interaction centers or virtual sites, that is, interactions between points (e.g. lone pairs) not located on atoms.

For solvents, especially water, a variety of molecular models is available, of which a few have been developed explicitly for use in mixed solute-water systems." 13491 Models for nonpolar solvents like carbon tetrachloride are also available.

When determining the parameters of the interaction function (1) there are essentially two routes to take. The most elegant procedure is to fit them to results (potential or field) of ab-initio quantum calculations on small molecular clusters. However, due to various serious approximations that have to be made in this type of procedure, the resulting force fields are in general not very satisfactory. The alternative is to fit the force field parameters to experimental data (crystal structure, energy and lattice dynamics, infrared, X-ray data on small molecules, liquid properties like density and enthalpy of vaporization, free energies of solvation, nuclear magnetic resonance data, etc.). In our opinion the best results have been obtained by this semi-empirical approach.

We wish to stress that one should fit force field parameters to properties of small molecules, which may be considered as building blocks of larger molecules such as proteins and DNA, and subsequently apply them to these larger molecules as a test without any further adaptations to improve the test results. The choice of a particular force field should depend on the type of system for which it has been designed. The MM2 force field is based on gas phase structures of small organic compounds.

The AMBERand CHARMM force fields are aimed at a description of isolated polypeptides and polynucieotides, in which the absence of a solvent (aqueous) environment is compensated by the use of a distance-dependent dielectric constant ε_r. The ECEPP and UNICEPP force fields use $\varepsilon_r = 4$, whereas the GROMOS I 2% force field uses $\varepsilon_r = 1$, since it has been set up for simulation of biomolecules in aqueous environment. This also holds for the OPLS. force field which is aimed at a proper description of solvation properties.

Force fields that are applicable to a more restricted range of compounds like ions liquid metals,

or carbohydrate, are manifold. The quality of the various force fields should be judged from the literature concerning their application to molecular systems.

Inclusion of Polarizability

The term in the interaction function (1) representing the nonbonded interactions consists only of a summation over all pair interactions in the system. Nonbonded many-body interactions are neglected. Yet, inclusion of polarizability of atoms or bonds will be inevitable if one would like to simulate, e.g., the binding of a charged hgdnd, which will polarize the part of the receptor to which it is binding (Table 1). About 10% (4 kJ mol[-1]) of the energy of liquid water is polarization energy. The infinite frequency dielectric constant of water is $\varepsilon_\infty = 5.3$, its effective dipole moment changes from 1.85 D in the gas phase to 2.4 D in the liquid. An analysis of the contribution of polarizability to local fields in proteins is given in Ref.

Inclusion of polarizability in the molecular model is not too difficult, as can be observed from the following simple outline. We consider a system of N point dipoles with Cartesian position vectors r_i, dipole moments p, and polarizabilities a, (constant, isotropic). The induced dipoles Δp_i obey a field equation (2), where E, denotes the electric field

$$\Delta p_i = \alpha_i E_i = \alpha_i \sum_{\substack{j=1 \\ \neq i}}^{N} T_{ij}[p_j + \Delta p_j]$$

(2)

at position r_i and the field tensor T_{ij} is given by Equation (3). The field equation can be solved for Δp_i either by the inversion

$$T_{ij} = (4\pi\varepsilon_0)^{-1} [3\, r_i r_j / r_{ij}^2 - 1]\, r_{ij}^{-3}$$

(3)

of a matrix of size 3N x 3N, or by iteration. The former method is impractical for a molecular system containing thousands of atoms. The latter method is well suited for use in MD simuiations, since the induced dipoles at the previous MD integration time step, $\Delta p_i(t - \Delta t)_i$, will be an excellent starting point for the iterative solution of the field equation (2) at time t, which yields Api(t). In this way inclusion of polarizability may increase simulation times by only 20-100%.

Computer simulations that include polarizability have only been performed for a limited number of molecular systems, such as ionic liquids, and a single protein. Many aspects of the treatment of polarizability in MD simulations are still under investigation. What is the most practical way to model polarizability?

1. By inducing point dipoles, as sketched above,
2. by changing the magnitudes of (atomic) charges,
3. by changing the positions of (atomic) charges.

How should the size of the atomic polarizabilities α_i be chosen, etc?

Treatment of Long Range Coulomb Forces

The summation of the last term in the interaction function (3) covering the nonbonded interaction runs over all atom pairs in the molecular system. It is proportional to N_2, the square of the number of atoms in the system. Since the other parts of the calculation are proportional to N, computational efficiency can be much improved by a reduction of this summation.The simplest procedure is to apply a cut-off criterion for the non-bonded interaction and to use a list of neighbor atoms lying within the cut-off, which is only updated every so many simulation steps. The cut-off radius R, usually has a value between 6 8, and 9 8, and the neighbor list is updated about every 10 or 20 MD time steps. This procedure does not introduce any errors as long as the range of the nonbonded interaction is smaller than R_c. However, the Coulomb term in (1) is proportional to r^1, which makes it long-ranged. If the molecular model does not involve bare (partial) charges on atoms, but only dipoles or higher multipoles, the electrostatic interaction term becomes proportional to r^3, which makes it of much shorter range. However, when dipoles are correlated over larger distances, as is the case for secondary structure elements like α-helices in proteins, their interactions again become long-ranged. In the following subsections we briefly discuss a variety of methods for the treatment of long-range electrostatic interactions in molecular systems.

Distance Dependent Dielectric Constant

A simple way to reduce the range of the Coulomb interaction is to introduce a relative dielectric constant proportional to r, viz. $\varepsilon_r = r$ (in $\mathring{A}$). The interaction becomes proportional to r-'. It is difficult to find a physical argument in favor of this approximation. Its overall effect is to effectively reduce all types of long-ranged interactions. Due to its simplicity it has been incorporated into a number of current force fields. Nevertheless, we think this approximation is too crude for practical applications.

Cut-off Radius and Neutral Groups of Atoms

When applying a cut-off radius &, the discontinuity of the interaction at a distance $r = R_c$, will act as a noise source in a MD simulation, and this will artificially increase the kinetic energy of the atoms and thus the temperature of the system. A possible way to reduce the noise is to multiply the nonbonded interaction term in (1) with a so-called switching function (4), which satisfies the conditions $S(R_S) = 1$,

$$S(r) = \begin{cases} 1 & r < R_S \\ (R_C - r)^2 (R_C + 2r - 3R_S)/(R_C - R_S)^3 & R_S < r < R_C \\ 0 & r > R_C \end{cases}$$

(4)

$$dS/dr(R_S) = 0, \quad S(R_C) = 0, \quad dS/dr(R_C) = 0.$$

Its effect is to smoothen the interaction on the interval (R_s, R_c), but there is no physical argument for its use. An empirical evaluation of the use of switching functions can be found in Ref. When the (partial) atomic charges of a group of atoms add up to exactly zero, the leading term of the electric interaction between two such groups of atoms is of dipolar character, that is, proportional to r^{-1}. For larger r the sum of the rmonopole contributions of the various atom pairs to the group group interaction will become zero. Therefore, the range of the electric interaction can be considerably reduced when atoms are assembled in neutral groups, so-called charge groups, which have a zero net charge, and for which the electric interaction with other (groups of) atoms is either calculated for all atoms of the charge group or for none. When using the charge group concept the cut-off criterion should be applied to the distance between groups, and a switching function like (4) must nor be used, since it distorts in the interval (R_s, R_c) the proper r-l weighting of atom-atom monopole interactions.

Cut-off' Radius plus Multipole Expansion

The technique of using neutral groups of atoms is based on the more general fact that a charge distribution of a finite group of atoms can be approximated by a multipole expansion (monopole, dipole, quadrupole, etc.). The electric interaction between two groups of atoms can be formulated as the product of the two multipole expansions. The terms in the resulting expression can be grouped according to their distance dependence: monopole-monopole (r^{-1}), monopoleedipole (r^{-2}), monopolequadrupole and dipole-dipole (r^{-3}), etc. At long distance only the leading terms in the series need to be taken into account. Application of this multipole expansion approximation in MD simulations was suggested by Ladd.". It is also used in combination with the "cut-off plus atom pair list" technique in Ref.

Twin Range Method

The twin range method is illustrated in Figure 7. Two cut-off radii are used. The atomsjlying within a distance R_f from atom i are stored in a neighbor list of atom i. The interaction of the atoms j for which $R_c^1 < r_{ij} < R_c^2$; with atom i, is stored in the form of a so-called long-range force Ff.' on atom i. At each MD time step the nonbonded interaction consists of two contributions: (I) the short-range part which is calculated from the neighbor list using the actual atom positions, and (2) the long range part F^{lr} which is kept fixed during N; time steps. Neighbor list and long range force F^{lr} are simultaneously updated every N_c^1 $(10-100)$ time steps. This twin range method is based on the assumption that the high-frequency components of the long range force may be safely neglected. For example, the mean and low frequency field of the correlated peptide dipoles of the long a-helix are accurately accounted for, only the fast $(\lesssim 0.2 \text{ ps})$ vibrations are neglected (see Fig. 7). The twin range method can also be applied using charged groups instead of atoms.

Continuum Approximations to the Reaction Field

In the previous subsections approximations to the long-range interaction between atoms were discussed that are explicitly taken into account as degrees of freedom in the system that is simulated

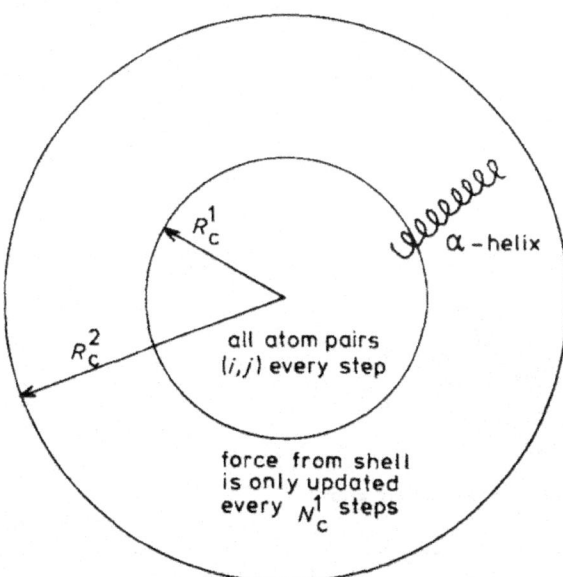

Figure 7: Twin range method. High-frequency components of the force on the central atom exerted by atoms between @ and R: are neglected.

If parts of the system are homogeneous, like the bulk solvent surrounding a solute, the number of atoms or degrees of freedom can be reduced considerably by modeling of the homogeneous part as a continuous medium, e.g. a continuous dielectric. In this type of approximation the system is divided into two parts: (1) an inner region where the atomic charges q, are explicitty treated (dielectric constant ε_1,), and (2) an outer region which is treated as a continuous medium with dielectric constant ε_2 and ionic strength I (see Fig. 8). The potential in the inner region

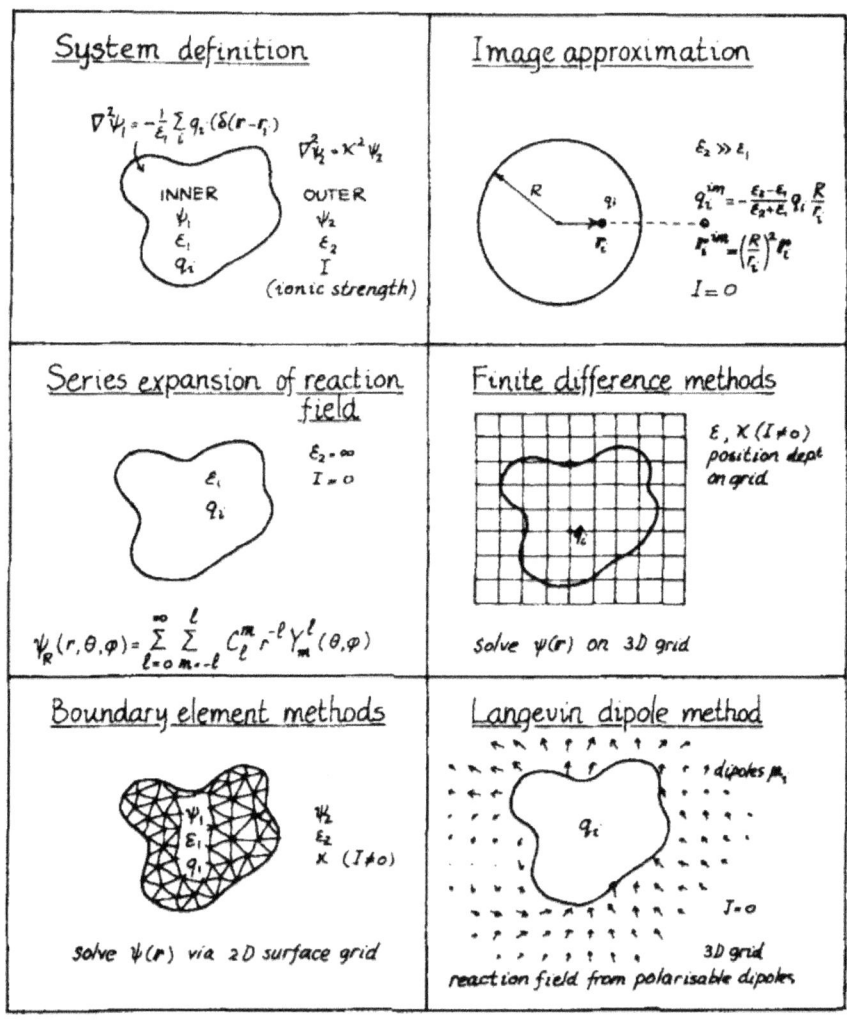

Figure 8: Non-periodic methods for computing long-range Coulomb forces.

$$\psi_1(r) = \psi_C(r) + \psi_R(r)$$

$$(5)$$

Eq. (5). consists of two terms. The first one

Eq. (6). is a direct Coulomb term due to the charges in the inner region,

$$\psi_C(r) = (4\pi\varepsilon_0\varepsilon_1)^{-1} \sum_{i=1}^{N} q_i/|r - r_i|$$

$$(6)$$

which is a solution to the Poisson equation (7) for the inner

$$\vec{\nabla}^2\psi(r) = -\varepsilon_1^{-1} \sum_{i=1}^{N} q_i\delta(r - r_i)$$

$$(7)$$

region. The second term is the reaction field potential $\psi_R(r)$, which satisfies the Poisson equation with zero source terms (no charges), that is, the Laplace equation (8). This means

$$\vec{\nabla}^2\psi(r) = 0.$$

$$(8)$$

that $\psi_1(r)$ is also a solution of (7). The potential in the outer region is denoted by $\psi_2(r)$. If the outer region continuum has a finite ionic strength I, the potential $\psi_2(r)$ must be a solution of the Poisson-Boltzmann equation (9) with inverse Debye

$$\vec{\nabla}^2\psi(r) = \kappa^2\psi(r)$$

$$(9)$$

length $\kappa = 2IF^2/(\varepsilon_2 RT)$ in which Fis Faraday's constant, R is the gas constant, and T the temperature. For zero ionic strength, (9) reduces to (8). The boundary condition at infinity is given by (lo), and at the boundary between the inner

$$\lim_{r \to \infty} \psi_2(r) = 0$$

$$(10)$$

and outer regions the potential $\psi(r)$

Eq. (1 I). and the dielectric displacement $\varepsilon\vec{\nabla}\psi(r)$

Eq. (12). must be continuous:

$$\psi_1(r) = \psi_2(r) \qquad\qquad r = \text{boundary}$$

$$(11)$$

$$\varepsilon_1\nabla_n\psi_1(r) = \varepsilon_2\nabla_n\psi_2(r) \quad r = \text{boundary}$$

$$(12)$$

where the component of the gradient normal to the boundary is denoted by V_n

The exact form of $\psi(r)$ will depend on the shape of the boundary, the values of ε_2 and $\kappa_{,}$, and the computational method that is used to solve the field equations (7)-(12).

Continuum Methods: Image Charge Approximation to the Reaction Field

When the boundary is a sphere of radius R the reaction field due to a charge q_i located at position r_i with respect to the origin of the sphere can be approximated (dipolar approximation, $\varepsilon_2 \gg \varepsilon_1$) by the field that is generated by a socalled image charge $q_i^{im} = -(\varepsilon_2 - \varepsilon_1)(\varepsilon_2 + \varepsilon_1)q_i R/r_i$, that is located at position $r_i^{im} = (R/r_i)^2 r_i^{[711]}$ (see Fig. 8). The name of this approximation is due to the mirror type of behavior of the spherical boundary: if qi approaches the boundary, q_i^{im} will do likewise, causing a pole in the potential at the boundary. This means that the image approximation breaks down for charges close to the boundary. The other limitations of the method are the requirement of a spherical boundary, and the conditions that $I = 0$ and $\varepsilon_2 \gg \varepsilon_1$.

Continuum Methods: Series Expansion of the Reaction Field

When the boundary is of irregular shape the reaction field potential cannot be written in a closed form, but must be approximated numerically. If we assume that I = 0 and $\varepsilon_2 = \infty$, the following approach is possible. Since the reaction field potential $\psi_R(r)$) must satisfy the Laplace equation (8) it may be expanded in Legendre polynomials, which also satisfy Equation (8)

Eq. (13).. $\varepsilon_2 = \infty$, we have $\psi_2(r) = 0$

$$\psi_R(r, \theta, \varphi) = \sum_{l=0}^{\infty} \sum_{m=-l}^{+l} C_l^m r^{-l} Y_m^l(\theta, \varphi).$$
(13)

at the boundary, or using (11) and (5) Equation (14), at the boundary, with $\psi_c(r)$ fixed by (6). The expansion coefficients

$$\psi_R(r) = -\psi_c(r)$$
(14)

C_l^m " can be determined numerically by performing a least squares fit of $\psi_R(r)$ to the given $-\psi_c(r)$ at a chosen number of points on the boundary.

The method works (Berendsen and Zwinderman, private communication), but is rather expensive due to the matrix inversion inherent in the least squares fit procedure. Close to the boundary the approximation is poor. The other limitations are that $l = 0$ and $\varepsilon_2 = \infty$

Continuum Methods: Three-dimensional Finite Difference Techniques

Another numerical approach is to compute the electric potential on a three-dimensional (3D) grid using finite difference methods (Fig. 8). The atomic charges qi are distributed over grid points. To each grid point a position-dependent dielectric constant ε_1 is assigned. Then the field equations (7)-(12) can be numerically solved. This involves the solution of a set of linear equations for the potential at grid points. The method is too time-consuming to be used in MD simulations, since the density of grid points must be sufficient to adequately represent the atomic charge distribution. The use of a position-dependent dielectric constant introduces an arbitrary element into the method.

Continuum Methods: Two-dimensional Boundary Element Techniques

A recent development is the application of Green's function techniques to the electric field problem. The Green's

$$G_1(r|r_0) = \frac{1}{4\pi|r - r_0|}$$

(15)

function (15) for the inner region satisfies Equation (16) (a Poisson equation with charge ε_0 at r_0.). The Green's function for the outer region

$$\vec{\nabla}^2 G_1(r|r_0) = -\delta(r - r_0)$$

(16)

Eq. (17). satisfies Equation (18) (a Poisson-Boltzmann equation with charge ε_0 at r_0.).

$$G_2(r|r_0) = \frac{e^{-\kappa|r-r_0|}}{4\pi|r - r_0|}$$

(17)

$$\vec{\nabla}^2 G(r|r_0) = \varkappa^2 G(r|r_0) - \delta(r - r_0)$$

(18)

The next step is to integrate the following integrand over the

$$[G_1 \times \text{expression (7) for } \psi_1 - \psi_1 \times \text{expression (16) for } G_1]$$

(19)

inner region and to convert it by Green's theorem into an integral over the surface of the boundary region. This yields Equation (20), consisting of a surface term and a source term.

$$\psi_1(r) = \int_{surface} [G_1(r'|r)\nabla_n\psi_1(r') - \psi_1(r')\nabla_n G_1(r'|r)]d\sigma$$

$$+ \sum_{i=1}^{N} \frac{q_i}{4\pi\varepsilon_0|r - r_i|}$$

(20)

$$[G_2 \times \text{expression (9) for } \psi_2 - \psi_2 \times \text{expression (18) for } G_2]$$

(21)

$$\psi_2(r) = \int_{surface} [G_2(r'|r)\nabla_n\psi_2(r') - \psi_2(r')\nabla_n G_2(r'|r)]d\sigma$$

(22)

Application of the boundary conditions (11) and (12) yields two linear integral equations in ψ_1 and $\nabla_n\psi_1$ on the boundary surface. These equations can be solved numerically by discretization on a surface net, and subsequently solved by matrix inversion. The (large) matrix that is to be inverted only depends on the shape and position of the boundary surface, not on the positions of the charges in the inner region. This means that in a MD simulation the time-consuming matrix inversion can be avoided as long as the boundary remains fixed.

The promise of this method lies in the reduction of a 3D problem to a 2D problem. On the boundary surface a charge density is generated which produces the proper reaction field. Also the methods of Shaw and of Zauhar and Morgan transform the problem from a three-dimensional one into a two-dimensional one.

Langevin Dipole Model

The Langevin dipole model of Warshel uses a 3Dgrid, but it is not a continuum method. The medium in the outer region is mimicked by polarizable point dipoles pi on the grid points i of a 3D grid. The dipoles may be thought of as representing the molecular dipoles of the solvent molecules forming the outer region, so the spacing of the grid points corresponds to the size of the solvent molecules. The size and direction of a dipole p, is determined by the electric field E_i at the grid point according to the Langevin formula Eq. (23).

$$\mu_i = \mu_0 \left[\frac{e^{+C\mu_0 E_i/kT} + e^{-C\mu_0 E_i/kT}}{e^{+C\mu_0 E_i/kT} - e^{-C\mu_0 E_i/kT}} - \frac{1}{C\mu_0 E_i/kT} \right] \frac{E_i}{E_i}$$

(23)

Here, pa is the magnitude of the dipole moment of a solvent molecule, and Cis a parameter representing the molecular resistance to reorientation. The model has been applied to protein simulations.t641 Questionable features of this model are the representation of the field in the outer region and the discreteness near the boundary.

Periodic Lattice Summation Methods

In computer simulations of liquids or solutions, periodic boundary conditions are often used to minimize the boundary effects. The computational box containing the molecular system is surrounded by an infinite number of copies of itself (see Fig. 9). In this way an infinite periodic system is simulated. The electric interaction in a periodic system is obtained by a summation over all atom pairs in the central box (Fig. 9) and over all atom pairs of which one atom lies in the central box and the other is a periodic image Eq. (24).

$$E = (4\pi\varepsilon_0)^{-1} \frac{1}{2} \sum_{\substack{n=0 \\ i \neq j \text{ if } n=0}}^{\infty} \sum_{i=1}^{N} \sum_{j=1}^{N} \frac{q_i q_j}{|r_{ij} + n|}$$

(24)

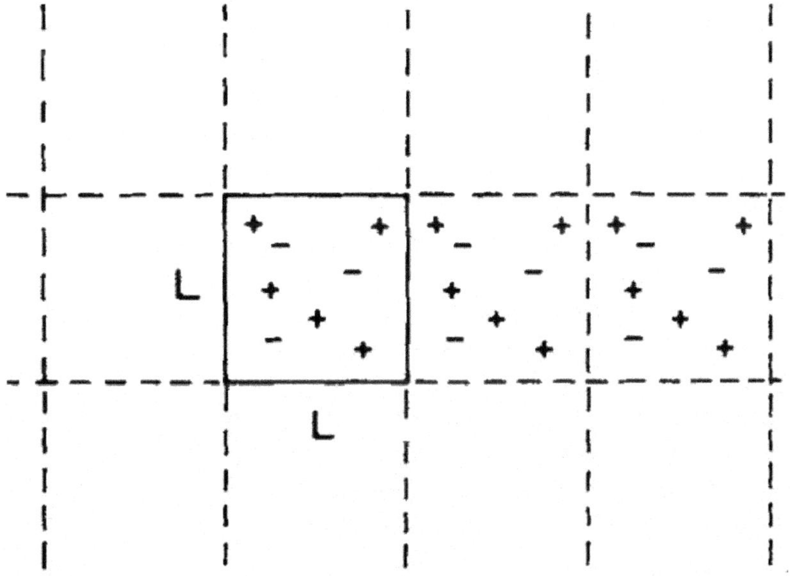

Figure 9: Periodic cubic system. Central computational box (solid line) with two of the infinite number of periodic image boxes.

The sum over n is a summation over all simple cubic lattice points $n = (n_x L, n_y L, n_z L)$ with n_x, n_y, n_z, integers. The infinite sum of point charges must be converted into a form which converges faster than Equation (24)

Periodic Methods: the Ewald Sum

Periodic Methods: the Ewald Sum The Ewald sum is a technique for computing the interaction of a charge and all its periodic images.16*76-781 The charge distribution e(r) in the system is an infinite set of point charges, mathematically represented as delta functions.

$$\varrho_i(r) = q_i \delta(r - r_i) \tag{25}$$

Each point charge q, is now surrounded by an isotropic Gaussian charge distribution of equal magnitude and opposite sign Eq. (26).

$$\varrho_i^G(r) = - q_i(\alpha/\sqrt{\pi})^3 e^{-\alpha^2|r - r_i|^2} \tag{26}$$

This smeared charge screens the interaction between the point charges, so that the interaction calculated using the screened charge distribution (27) becomes short-ranged

Eq. (28). due to the appearance of the complementary error function (29).

$$\varrho_i^s(r) \equiv \varrho_i(r) + \varrho_i^G(r) \tag{27}$$

$$E^s = (4\pi\varepsilon_0)^{-1} \frac{1}{2} \sum_{n=0}^{\infty} \sum_{i=1}^{N} \sum_{\substack{j=1 \\ i \neq j \text{ if } n=0}}^{N} \frac{q_i q_j}{|r_{ij} + n|} \text{erfc}(\alpha|r_{ij} + n|) \tag{28}$$

$$\text{erfc}(x) = 2\pi^{-1/2} \int_x^{\infty} e^{-y^2} dy \tag{29}$$

Thus, E^s can be well approximated using a finite summation in (28). Of course a purely Gaussian charge distribution $-\varrho_i^G(r)$ must be added to $\varrho_i^s(r)$ in order to recover the original charge distribution $\varrho_i(r)$. The interaction of these Gaussian distributions is expressed as a lattice sum in reciprocal space minus a self term

Eq. (30)., where

$$E^G = (4\pi\varepsilon_0)^{-1} 2\pi L^{-3} \sum_{\substack{k \neq 0}}^{\infty} k^{-2} e^{-k^2/(4\alpha^2)} \left| \sum_{j=1}^{N} q_j e^{-ik \cdot r_j} \right|^2$$

$$- (4\pi\varepsilon_0)^{-1} \pi^{-1/2} \alpha \sum_{j=1}^{N} q_j^2$$

(30)

$k = 2\pi L^{-1}(l_x, l_y, l_z)$ and l_x, l_y, l_z are integers. Due to the presence of the exponential factor the infinite lattice sum can be well approximated by a finite one. The parameter x should be chosen such as to optimize the convergence properties of both sums (28) and (30) which contribute to

$$E = E^s + E^G$$

The Ewald sum technique is routinely used in simulations of ionic systems. When applied to non-crystalline systems such as liquids or solutions it has the disadvantage of enhancing the artifact of the application of periodic boundary conditions.

Periodic Methods: Fourier Techniques

The Poisson equation (31) for the potential ψ(r) is a sec

$$\vec{\nabla}^2 \psi(r) = - \varepsilon_0^{-1} \varrho(r)$$

(31)

Here, the 3D Fourier transformation is defined according to Equation (34), and likewise for the charge density e(r). The inverse transform of Equation (33) yields (35).

$$\hat{\psi}(k) \equiv F\{\psi(r)\} \equiv (2\pi)^{-3/2} \int \psi(r) e^{-ik \cdot r} dr$$

(32)

$$\psi(r) = F^{-1}\{(\varepsilon_0 k^2)^{-1} F\{\varrho(r)\}\}$$

(33)

Due to the availability of fast Fourier transform techniques Equation (35) is a fast method to solve the Poisson equation on a periodic three-dimensional grid. For a practical implementation of the method we refer to Ref. It has been employed in the simulation of salts.

Summary

The choice of molecular model and force field is essential to a proper prediction of the properties of a system. Therefore, it is of great importance to be aware of the fundamental assumptions, simplifications and approximations that are implicit in the various types of models used in the literature. When treating molecular systems in which Coulomb forces play a role one should be aware of the way these long-ranged forces are handled. Since these forces play a dominant role in many molecular systems we have tried to give an overview of the different methods-and so the different approximations-that are in use. From Table I and the discussions in the previous sections it should be clear that a "best" force field does nor exist. What will be the best choice of model and force field will depend on the type of molecular system and the type of property one is interested in. This means that the molecular modeler must have a picture of the strengths and weaknesses of the variety of force fields that are available in order to make a proper choice.

Searching Configuration Space and Generating an Ensemble

Once the molecular model and force field V(r) have been chosen, a method to search configuration space for configurations with low energy V(r) has to be selected. Various methods are available, each with their particular strength and weakness, which depend on

- the form and type of the interaction energy function V(r),
- the number of degrees of freedom (size of the system),
- the type of degrees of freedom, viz. Cartesian coordinates

versus other coordinates (e.g. bond lengths, bond angles, torsional angles plus center of mass coordinates of a molecule).

Systematic Search Methods

If the molecular system contains only a small number of degrees of freedom (coordinates) and if V(r) does not have too many (relevant) minima upon variation of the degrees of freedom, it is possible to systematically scan the complete configuration space of the system. For example, one can describe conformations of n-alkanes in terms of C-C-C-C torsional angles, each of which has three conformational minima, trans (OO), gauche' (120") and gauche⁻(-120°). To find the lowest energy conformation of n-decane (seven torsional angles) one would have to compute V(r) for $3^7 = 21\ 87$ combinations of torsional angles. The relative weight of the different conformations in the ensemble representing this molecule at temperature T is given by the Boltzmann factor (36).

$$e^{-V(r)/k_B T} \qquad (36)$$

The computing effort required by a systematic search of the degrees of freedom of a system grows exponentially with their number. Only very small molecular systems can be treated by systematic search methods. The number of degrees of freedom that still can be handled within a reasonable computing time strongly depends on the complexity of the function V(r), that is, the time required to compute V(v) for each configuration. A possibility for speeding up the calculation is to split the selection of low V(r) configurations into different stages. In a first stage a simplified low computing cost energy function $V_{Simple}(r)$ is used to quickly discard high $V_{simple}(i)$ configurations. In a second stage the complete V(r) is only to be evaluated for the remaining configurations. This type of filtering procedure has been used to predict the loop structure in proteins and to predict the stable conformation of small peptides (five amino acid residues) for example. The basic problem of filtering using simplified forms of V(r) is to ensure that the simplified $V_{simple}(r)$ is a correct projection of the complete function V(r): when $V_{simple}(r)$ is large, V(v) should also be large, otherwise a configuration with low energy V(r) might be discarded in the first stage due to a high energy $V_{simpl}(r)$

Random Search Methods

If a system contains too many degrees of freedom, straightforward scanning of the complete configuration space is impossible. In that case a collection of configurations can be generated by random sampling. Such a collection becomes an ensemble of configurations when each configuration is given its Boltzmann factor

cf. Eq. (36). as weight factor in the collection. Two types of random search methods can be distinguished :

- Monte Carlo methods, in which a sequence of configurations is generated by an algorithm which ensures that the occurrence of configurations is proportional to their Boltzmann factors Eq. (36)..

- Other methods which produce an arbitrary (non-Boltzmann) collection of configurations, which can be transformed into a Boltzmann ensemble by applying the weight factors from Equation (36).

Monte Carlo Sirnulalion

The Monte Carlo (MC) simulation procedure by which a (canonical) ensemble is produced consists of the following steps.

Given a starting configuration r, a new configuration $r_{s+1} = r_s + \Delta r$ is generated by random displacement of one (or more) atoms. The random displacements Δr should be such that in the limit of a large number of successive displacements the available Cartesian space of all atoms is uniformly sampled. This does not mean that the actual sampling must be carried out in Cartesian space. It can be done, e.g., in internal coordinate space (r, 0, cp), but since the equivalent volume element is $r^2 \sin 0 \, dr d0 d\varphi$, the sampling in internal coordinate space must be non-uniform in order to produce a uniform sampling in terms of Cartesian coordinates.

The newly generated configuration r_{s+1} is either accepted or rejected on the basis of an energy criterion involving the change $\Delta E = V(r_{s+1}) - V(r_s)$ of the potential energy with respect to the previous Configuration. The new configuration is accepted when $\Delta E \leq 0$, or if $\Delta E > 0$ when $e^{-\Delta E/k_B T} > R$, where R, where R is a random number taken from a uniform distribution over the interval (0,1). Upon acceptance, the new configuration is counted and used as a starting point for the next random displacement. If the criteria are not met, the new configuration r,, is rejected. This implies that the previous configuration r, is counted again and used as a starting point for another random displacement.

It is relatively easy to understand that this procedure will generate a Boltzmann ensemble. We consider two configurations r_1 and r_2 with energies $E_1 = V(r_1) < V(r_2) = E_2$. The probability of stepping from configuration r_2 to r_1 equals 1, the reverse step has a probability $\exp(-(E_2 - E_1)/k_B T)$. When the populations p_1 and p_2 of the two configurations are in equilibrium, one has detailed balance conditions (37) or (38).

$$p_2 \cdot 1 = p_1 \cdot e^{-(E_2 - E_1)/k_B T} \tag{31}$$

$$p_1/p_2 = e^{-E_1/k_B T}/e^{-E_2/k_B T} \tag{32}$$

Each configuration occurs with a probability proportional to its Boltzmann factor Eq. (36).

The advantage of this (Metropolis) Monte Carlo or Boltzmann sampling over random sampling is that most sampled configurations are relevant (low energy), while with random sampling much computational effort is likely to be spent on irrelevant (high energy) configurations. In order to obtain high computational efficiency, one would like to combine a large (random) step size with a high acceptance ratio. This is possible when applying MC techniques to simulate simple atomic or molecular liquids.C_6. However, for complex systems involving many covalently bound atoms, a reasonable acceptance ratio can only be obtained for very small step size. This is due to the fact that a random displacement will inevitably generate a very high bond energy of the bonds of the displaced atom. This makes MC methods rather inefficient for (macro) molecular systems.

Distance Geometry Methods

Distance geometry (DG) is a general method for converting a set of bounds on distances between atoms into a configuration of these atoms that is consistent with these bounds. The emergence of 2D-NMR techniques has spurred a renewed interest in techniques to obtain three-dimensional molecular structures from atom-atom distance information.The DG method is also applied in pharmacophore modelingrs. or enzyme substrate docking .I M. Blaney, private communication. to generate a collection of ligand structures compatible with a set of atom -atom distance bounds.

In distance geometry a molecular structure is described in terms of the set of all pairwise interatomic distances, which can be written in the form of a so-called (symmetric) distance matrix. By entering maximum distances between atoms of pairs in the upper right-hand triangle, and minimum distances in the lower left-hand triangle of the distance matrix it becomes a distance bounds matrix. Such a matrix describes the complete configuration space accessible to the molecule within the specified bounds. In a distance geometry calculation a set of random configurations is generated by choosing atom-atom distances at random within the specified bounds, and subsequently converting the resulting distance matrix into a structure in three-dimensional Cartesian space using a so-called embedding algorith.

The DG method is a powerful method for generating a set of configurations compatible with a set of atom-atom distances, but also has a number of limitations. It is not possible to apply an energy function $V(r)$ like Equation (1) in a DG calculation. The energy function has to be converted into a function of atom-atom distances only, and subsequently it must be simplified to a set of bounds on these distances, by which procedure much of the information present in $V(r)$ will be lost. As a consequence the DG method cannot properly handle solvent configurations, since a limited-distance description of a liquid lacks the ability to give proper statistical weight to the great variety of possible configurations. Another problem is the characterization of the distribution of generated three-dimensional configurations. Since the conversion from distance space into three-dimensional Cartesian space is non-linear, a uniform sampling of distances between the lower and upper bounds will certainly not produce a set of configurations uniformly distributed in Cartesian space.

Other Random Search Technique

There exist infinitely many ways for generating a set of molecular configurations in a random manner. Whether the set of generated configurations may be considered to form an ensemble that can be used for statistical mechanical evaluation of quantities will depend on the sampling properties of the method that is used. It should sample the important (low energy) regions of configuration space and the configurations should occur according to their Boltzmann weight factors. Otherwise the set of configurations can only be viewed as such, not as an ensemble representative of the state of the system that is considered.

Dynamic Simulation Methods

Another way to generate an ensemble of configurations is to apply Nature's laws of motion for the atoms of a molecular system. This has the additional advantage that dynamical information about the system is obtained as well. The two major simulation techniques of this type are Molecular Dynamics (MD), in which Newton's equations of motion are integrated over time, and Stochastic Dynamics (SD), in which the Langevin equation of motion for Brownian motion is integrated over time.

Molecular Dynamics Simulation

In the Molecular Dynamics (MD) method a trajectory (configurations as a function of time) of the molecular system is generated by simultaneous integration of Newton's equations of motion (39) and (40) for all the atoms in the

$$d^2r_i(t)/dt^2 = m_i^{-1} F_i \qquad (39)$$

$$F_i = - \partial V(r_i, \ldots r_N)/\partial r_i \qquad (40)$$

system. The force on atom i is denoted by 6 and time is denoted by t. MD simulation requires calculation of the gradient of the potential energy V(r), which therefore must be a differentiable function of the atomic coordinates ri. The integration of Equation (39) is performed in small time steps At, typically 1-10 fs for molecular systems. Static equilibrium quantities can be obtained by averaging over the trajectory, which must be of sufficient length to form a representative ensemble of the state of the system. In addition, dynamic information can be extracted. Another asset of MD simulation is that non-equilibrium properties can be efficiently studied by keeping the system in a steady non-equilibrium state, as is discussed in Section 5.4. Viewed as a technique to search configuration space, the power of MD lies in the fact that the kinetic energy present in the system allows it to surmount energy barriers that are of order of k, Tper degree of freedom. By raising the temperature T a larger part of conformation space can be searched, as has been shown by DiNola et al.,'881 who generated a series of different conformations of the hormone somatostatin by applying MD at T = 600 K and at 1200 K. At the elevated temperature the total energy and potential energy are monitored for conspicuous fluctuations which may signal a possibly significant conformational change. When minima in the total energy occur, the system is cooled down and equilibrated at normal temperature (300 K). In this way different conformations with comparable free energy were obtained. We note however that the search at elevated temperature favors selection of higher entropy conformations.

Searching conformation space by MD is expected to be efficient for molecules up to about 100 atoms. For larger molecules, which may and are likely to show a particular topological fold, MD methods will not be able to generate major topological rearrangements. Even when the barriers separating two topologically different low energy regions of conformation space are of the order of k,T, the time needed for traversing them may be much too long to be covered in a MD simulation of 10-100 ps.

Stochastic Dynamics Simulation

The method of stochastic dynamics (SD) is an extension of MD. A trajectory of the molecular system is generated by integration of the stochastic Langevin equation of motion (41 1.

$$d^2 r_i(t)/dt^2 = m_i^{-1} F_i + m_i^{-1} R_i - \gamma_i dr_i(t)/dt \tag{41}$$

Two terms are added to Equation (39), a stochastic force R_i and a frictional force proportional to a friction coefficient γ_i. The stochastic term introduces energy, the frictional term removes (kinetic) energy from the system, the condition for zero energy loss being that given in Equation (42), where

$$\langle R_i^2 \rangle = 6 m_i \gamma_i k_B T_{ref} \tag{42}$$

T_{erf} is the reference temperature of the system SD can be applied to establish a coupling of the individual atom motion to a heat bath, or to mimic a solvent effect.lgO. In the latter case the stochastic term represents collisions of solute atoms with solvent molecules and the frictional term represents the drag exerted by the solvent on the solute atom motion. An introduction to SD simulation techniques is given in Refs.

Other Search Methods

There exists a variety of other methods for searching configuration space which are variants of the basic types discussed in the previous subsections.

The method of simulated annealing is a MC or a MD simulation in which the temperature is gradually lowered to OK. In this way the system generally ends up in a lower energy state than when an ordinary gradient energy minimizer is used.

In a MD simulation the atomic velocities are large when the system explores the low energy regions of the potential energy function V(r), and small when it crosses barriers at higher energy. For an efficient searching of configuration space one would like to invert this behavior: the atoms should move slowly in the valleys of V(r). A search algorithm developed along this line seems to perform well for certain systems.1941 Within the framework of MD the ability of atoms to cross barriers can be greatly enhanced by keeping their kinetic energy nearly constant by a tight coupling to a heat bath.

Another possibility is the combination of different techniques; Monte Carlo simulation, gradient minimization, etc. in one computational scheme.

Summary

The various methods for searching configuration space can be classified as follows:

1. Systematic search methods, which scan the complete configuration space of the molecular system
2. Methods which aim at generating a representative set of configurations. These may be divided into two types.

- Non-step methods, such as the DG method, which generate a (at least in principle) uncorrelated series of random configurations.
- Step methods, such as MC, MD and SD, which generate a new configuration from the previous one.

Most search techniques fall in this class. They can be distinguished by the way the step direction and the step size are chosen :

- according to the gradient $-\vec{\nabla}V(r)$,
- according to a memory of the path followed so far, or
- at random.

For example, in the MC method, the step direction is chosen at random, the actual step size is determined by the change in energy AE (gradient) and a random element in case AE > 0. In a MD simulation both step size and direction are determined by the force (gradient) and the velocity (memory). In general a search algorithm will make a step which is a linear combination of the gradient, previous steps (memory) and a random contribution. It will depend on the energy surface V(r) which is the optimal linear combination of these three ingredients.

Another classification of algorithms which generate a set of configurations is whether they produce an ensemble or not. Of the schemes discussed above only MC, MD and SD generate a Boltzmann weighted ensemble.

Boundary Conditions

When simulating a system of finite size, some thought must be given to the way the boundary of the system will be treated. The simplest choice is the vacuum boundary condition. When simulating a liquid, solution or solid rather than a molecule in the gas phase, it is common practice to minimize edge or wall effects by the application of periodic boundary conditions. If the irregularity of the system is incompatible with periodicity, edge or wall effects may be reduced by treating part of the system as an extended wall region in which the motion of the atoms is partially restricted.

Vacuum Boundary Condition

Simulation of a molecular system in vacuo, that is, without any wall or boundary, corresponds to the gas phase at zero pressure. When the vacuum boundary is used for a solid or a molecule in solution, properties of atoms near or at the surface of the system will be distorted.Ig7. The vacuum boundary condition may also distort the shape of a (nonspherical) molecule, since it generally tends to minimize the surface area. Moreover, the shielding effect of a solvent with high dielectric permittivity, like water, on the electric interaction between charges or dipoles in a molecule is lacking in vacuo. We note that the water molecules do not need to be positioned between the charges or dipoles in order to produce a screening of the interaction between these. Therefore, simulation of a charged extended molecule like DNA in vacuo is a precarious undertaking. Solvation of DNA in a sphere with solvent molecules will shift the boundary effects from the DNA-vacuum interface to the water-vacuum interface and so improve the treatment of the DNA.Ig8' The best results in vacuo are obtained for relatively large globular molecular systems.

Periodic Boundary Conditions

The classical way to minimize edge effects in a finite system is to use periodic boundary conditions. The atoms of the system that is to be simulated are put into a cubic, or more generally into any periodically space-filling shaped box, which is treated as if it is surrounded by 26 $(= 3^3 - 1^3)$ identical translated (over distances k $\pm R_{box}$ in the x-, y-, z-di--directions) images of itself. The next layer of neighbor images of the central computational box contains $5^3 - 3^3 - 1^3 = 98$ boxes, and so on. When an infinite lattice sum of the atomic interactions is to be performed (Sections 3.1.3.1 1 - 13), the interactions of an atom in the central computational box with all its periodic images are computed. In most cases this is not desirable. Then only interactions with nearest neighbors are taken into account. The black atom in the central computational box in Figure 10a will only interact with atoms or images of atoms that lie within the dashed line (nearest image, NI, or minimum image, MI, approximation). The anisotropy of the interaction due to the cubic shape of the nearest image box can be avoided by the application of a spherical cut-off (radius R_c). The periodic boundary condition affects not only the computation of the forces, but also the positions of the atoms. It is common practice (though not necessary) to keep the atoms together, that is, in the central computational box: when an atom leaves the central box on one side, it enters it with identical velocity at the opposite side at the translated image position.

Application of periodic boundary conditions means that in fact a crystal is simulated. For a liquid or solution the periodicity is an artifact of the computation, so the effects should be minimized. An atom should not simultaneously interact with another atom and a periodic image of that atom. Consequently the length R_{box} of the edge of the periodic box should exceed twice the cut-off radius R_c. Possible distorting effects of the periodic boundary conditation may be traced by simulation of a system in differently shaped boxes (see below) of different size.

a   b

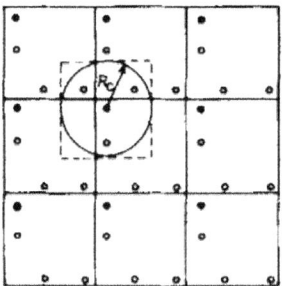
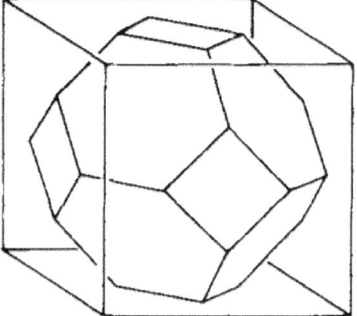

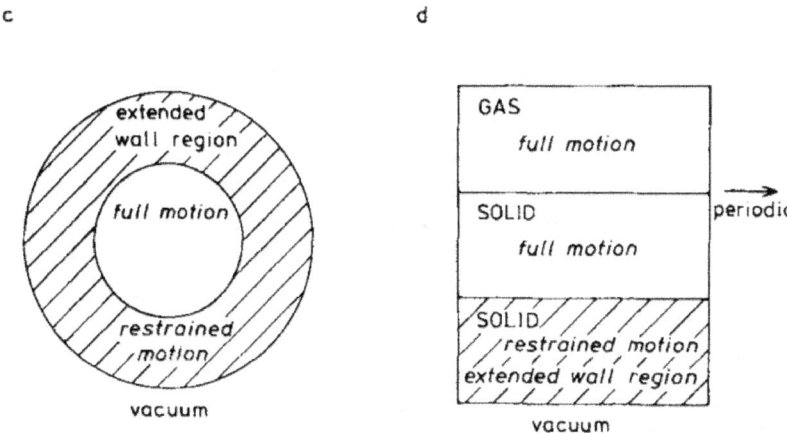

Figure 10: Types of boundary conditions: a) cubic periodic, b) troncated octahedron periodic, c) spherical extended wall region, d) flat extended wall region for a gas-solid simulation.

When simulating a spherical solute, use of a more spherically shaped computational box instead of a cubic or rectangular one may considerably reduce the number of solvent molecules that is needed to fill the remaining (after insertion of the solute) empty space in the box. A more spherically shaped space-filling periodic box is a truncated octahedron, (Fig. 10b).1103. It is obtained by cutting off the corners of a cube in such a way that the symmetry of the cube is main

tained, and such that the distance between opposite hexagonal planes equals $\frac{1}{2}\sqrt{3}$ times the distance between opposite square planes. In this way half the volume of the original cube is cut away. The ratio of the inscribed-sphere volume to total volume of the truncated octahedron is 0.68, whereas it is only 0.52 for a cube. Use of a truncated octahedron instead of a rectangular periodic box may yield a sizeable reduction of the system to be simulated. For example, the small protein bovine pancreatic trypsin inhibitor (BPTI) (58 amino acid residues) consists of 568 atoms (hydrogen atoms bound to carbons excluded), and its dimensions are 28 A by 28 A by 40 8,. If we approximate it by a sphere of radius 16 A and put it into a cube with a minimum solvent layer thickness of 6 8, to the walfs, about 2300 water molecules (6900 atoms) are needed to fill the non-protein part of the cube of volume $[2 \times (16 + 6)]^3 A^3$. To fill the non-protein part of a truncated octahedron of volume $\frac{1}{2}[4 \times (16 + 6)/\sqrt{3}]^3 \text{ Å}^3$ takes about 1600 water molecules (4800 atoms). This means a reduction of the system size by at least one quarter of the number of atoms.

Extended Wall Region Boundary Condition

If a molecular system is irregular, i.e. not nearly translationally periodic, periodic boundary conditions cannot be applied. In this case the distorting effect of the vacuum outside the molecular system may be reduced by designating a layer of atoms of the system to be an extended wall region, in which the motion of the atoms is restrained in order to avoid the deforming influence of the nearby vacuum. The atoms in the extended wall region can be kept fixed Is') or harmonically restrained to stationary positions. Their motion may be coupled to a heat bath, e.g. by applying stochastic dynamics in order to account for exchange of energy with the surroundings. In any case the type of force applied to these atoms should be chosen such that their motion in the finite system resembles as closely as possible the true motion in an infinite system. The extended wall region forms a buffer between the fully unrestrained part of the system and the (unrealistic) vacuum surrounding it. This extended wall region technique has been applied in the simulation of solid and protein Although the technique yields considerable savings in computing time, it should be carefully analyzed how far the restraining of the motion of the atoms in the wall region will affect the motion of the freely simulated atoms in the system.

Different Types of Molecular Dynamics

When Newton's equations of motion (39) and (40) are integrated the total energy is conserved (adiabatic system) and if the volume is held constant the simulation will generate a microcanonical ensemble. For various reasons this is not very convenient and a variety of approaches has appeared in the literature to yield a type of dynamics in which temperature and pressure are independent variables rather than derived properties. When MD is performed in nonequilibrium situations in order to study irreversible processes, catalytic events or transport properties, the need to impress external constraints or restraints on the system is apparent. In such cases the temperature should be controlled as well in order to absorb the dissipative heat produced by the irreversible process. But also in equilibrium simulations the automatic control of temperature and pressure as independent variables is very convenient. Slow temperature drifts that are an unavoidable result of force truncation errors are corrected, while also rapid transitions to new desired conditions of temperature and pressure are more easily accomplished.

Constant- Temperature Molecular Dynamics

Several methods for performing MD at constant temperature have been proposed, ranging from ad-hoc rescaling of atomic velocities in order to adjust the temperature, to consistent formulation in terms of modified Lagrangian equations of motion that force the dynamics to follow the desired temperature constraint. Different types of methods can be distinguished.

1.   Constraint methods, in which the current temperature T(t) at time point t is exactly equal to the desired reference temperature To. This can be achieved by rescaling the velocities at each MD time step by a factor $[T_0/T(t)]^{1/2}$, where the temperature T(t) is defined in terms of the kinetic energy through equipartition IEq. (43).. The number of degrees of freedom in the system is N_{df}.

$$E_{kin}(t) = \sum_{i=1}^{N} \frac{1}{2} m_i v_i^2(t) = \frac{1}{2} N_{df} k_B T(t)$$

(43)

The same result can be obtained in a more elegant way by a modification of the equations of motion (39); such that the kinetic energy instead of the total energy becomes a constant of motion. This method has two disadvantages. First, numerical inaccuracy of the algorithm may produce a drift in temperature that is not stabilized, since the reference temperature To does not appear in the equations to be integrated. Second, although the system is modeled by a Hamiltonian, this Hamiltonian does not represent a physical system, for which fluctuations in the kinetic energy are characteristic. In practical applications which are aimed at realistically simulating a physical system, the use of a nonphysical Hamiltonian is questionable, even though mathematically consistent equations of motion are obtained.

1.   Extended system methods,

I in which an extra degree of freedom s representing a heat bath is added to the atomic degrees of freedom of the molecular system. A kinetic and a potential energy term invoking this extra degree of freedom are added to the Hamiltonian and the simulation is carried out for the $N_{df} + 1$ degrees of freedom of this extended system. Due to the presence of a kinetic term $\frac{1}{2} m_s (ds/dt)^2$ in the Hamiltonian, energy is flowing dynamically from the heat bath to the system and back, the speed being controlled by the inertia parameter m_s. A disadvantage of this type of second-order coupling to a heat bath is the occurrence of spurious energy oscillations, the period of which depends on the value of the adjustable coupling parameter m_s.

$$dT(t)/dt = \tau_T^{-1}[T_0 - T(t)]$$

(44)

The kinetic energy can be changed by AEkim in a MD time step At by scaling all atomic velocities V, with a factor I. Using (43) we get (45).

$$\Delta E_{kin} = (\lambda^2 - 1)\tfrac{1}{2}N_{df}\,k_B\,T(t)$$

$$(45)$$

If the heat capacity per degree of freedom of the system is denoted by ctf the change in energy

Eq. (45). leads to a

$$\Delta T = [N_{df}c_V^{df}]^{-1}\,\Delta E_{kin}$$

$$(46)$$

change in temperature

Eq. (46)., which should be equal to AT as determined by (44). Solving Equations (44)-(46) for I we obtain equation (47).

$$\lambda = [1 + c_V^{df}(k_B/2)^{-1}\Delta t\,\tau_T^{-1}(T_0/T(t)-1)]^{1/2}$$

$$(47)$$

The heat capacity per degree of freedom c_V^{df} may not be accurately known for the system. This has no consequence for the dynamics since the temperature relaxation time τ_T is an adjustable parameter. This aperiodic coupling to a heat bath through a first-order process has the advantage over the extended system methods that the response to temperature changes is non-oscillatory and that it is rather easy to implement through a simple velocity scaling using Equation (47). The coupling can be chosen sufficiently weak (sufficiently large τ_T) to avoid disturbance of the system and sufficiently strong (small τ_T) to achieve the desired result.

Stochastic methods, in which the individual atomic velocities 5, are changed stochastically. Andemen" 14. proposed a Maxwellian re-thermalization procedure by stochastic collisions, in which the mean time between collisions plays the role of adjustable parameter determining the strength of the coupling to the heat bath. Heyes has suggested a Monte Carlo type technique for selection of new velocities. Others1116. used the Langevin equation (41) to achieve the coupling to a heat bath, the strength of the coupling being determined by the value of the atomic friction coefficients γ_i

Constant-Pressure Molecular Dynamics

For an isotropic system the pressure is a scalar defined by Equation (48), where V denotes the volume of the computational box and the virial E is defined as in Equation (49).

$$P = 2/(3\,V)[E_{kin} - \Xi]$$

$$(48)$$

$$\Xi = -\frac{1}{2} \sum_{\text{pairs } (i,j)}^{N} r_{ij} \cdot F_{ij}$$

(49)

Here, $r_{ij} = r_i - r_j$ and F_{ij} is the force on atom i due to atom j. For molecular systems, forces within a molecule may be omitted together with kinetic energy contributions of intramolecular degrees of freedom. A pressure change can be achieved by scaling of the volume of the box and by changing the virial through a scaling of interatomic distances.

The various methods for carrying out MD at constant pressure are based on the same principles as the constant temperature scheme with the role of the temperature played by the pressure and the role of the atomic velocities played by the atomic positions. The following methods can be distinguished.

Constraint methods," "I in which the equations of motion are modified such that the pressure instead of the volume becomes a constant of motion. This type of method has the same two disadvantages as its constant temperature counterpart.

Extended system methods, in ' which an extra degree of freedom V, the volume of the box, is added to the atomic degrees of freedom of the system. A kinetic energy term, ; $\frac{1}{2}m_v(dV/dt)^2$ ' and a potential energy term, PV, involving the extra degree of freedom are added to the Hamiltonian, and the simulation is carried out for the $N_{df}+1$ degrees of freedom of the extended system. The rate of volume change is governed by the inertia parameter my. A disadvantage of this second-order coupling method is the occurrence of spurious volume oscillations, the period of which depends on the size of the adjustable parameter m_v.

Weak coupling methods, in which the atomic equations of motion are modified such that the net result on the system is a first-order relaxation of the pressure towards a reference value P_o.

$$dP(t)/dt = \tau_p^{-1}[P_0 - P(t)]$$

(50)

Scaling the atomic coordinates ri and the edges of the computational box by a factor p leads to a volume change

$$\Delta V = (\mu^3 - 1)V$$

(51)

The pressure change ΔP due to this change in volume is given by Equation (52), where the isothermal compressibility of

$$\Delta P = -(\beta_T V)^{-1} \Delta V$$

(52)

the system is denoted by β_T. Solving Equations (50)-(52) for p we get Equation (53).

$$\mu = [1 - \beta_T \Delta t \tau_p^{-1} (P_0 - P(t))]^{1/3} \tag{53}$$

Since the pressure relaxation time τ_p is an adjustable parameter, an accurate value for the compressibility of the system is not required. This first-order pressure coupling method has advantages corresponding to those of the weak temperature coupling method cf. Eqs. (44)-(47).. The expressions (48)-(53) can easily be modified to apply to a general anisotropic triclinic system. Virial, kinetic energy, pressure, and the scaling factor p will become Cartesian tensors, and the volume V becomes the determinant of the matrix formed by the vectors a, b and c representing the edges of the computational box.

Stochastic methods for constant-pressure MD have not yet been proposed. They would involve random changes of the box volume.

Algorithms for Molecular and Stochastic Dynamics

Integration Schemes for Molecular Dynamics

Newton's equation of motion (39), a second-order differential equation, can be written as two first-order differential equations (54) and (55) for the particle positions $r_i(t)$ and

$$dv_i(t)]dt = m_i^{-1} F_i(\{r_i(t)\}) \tag{54}$$

$$dr_i(t)/dt = v_i(t) \tag{55}$$

velocities $v_i(t)$ respectively. The forces Fi are obtained from the potential energy function through Equation (40); they therefore depend on the configuration $\{r_i(t)\}$ of the system. A simple algorithm for integration of Equations (54) and (55) in small time steps Δt is obtained as follows.

$$v_i(t_n + \Delta t/2) = v_i(t_n) + dv_i/dt]_{t_n} \Delta t/2$$
$$+ d^2 v_i(t)/dt^2]_{t_n} (\Delta t/2)^2/2! + O(\Delta t^3)$$
$$v_i(t_n - \Delta t/2) = v_i(t_n) - dv_i/dt]_{t_n} \Delta t/2$$
$$+ d^2 v_i(t)/dt^2]_{t_n} (\Delta t/2)^2/2! + O(\Delta t^3)$$

$$v_i(t_n + \Delta t/2) = v_i(t_n - \Delta t/2) + m_i^{-1} F_i(\{r_i(t_n)\}) \Delta t + O(\Delta t^3) \tag{56}$$

Using the same procedure for Taylor expansions of $r_i(t)$ at time point $t = t_n + \Delta t/2$ and using Equation (55) we obtain Equation (57).

$$r_i(t_n + \Delta t) = r_i(t_n) + v_i(t_n + \Delta t/2)\,\Delta t + O(\Delta t^3).$$

(57)

Equations (56)-(57) form the so-called leap-frog scheme. Its name is illustrated in Figure 1 I. It is one of the most accurate, stable, and yet simple and efficient algorithms available for molecular dynamics of fluid-like systems. Our

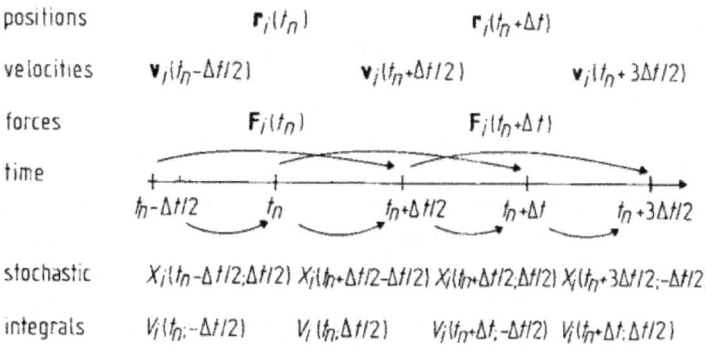

Figure 11: The leap-frog scheme for integration of Newton's equations of motion and Langevin's equations of motion.

preference for using the leap-frog scheme instead of other algorithms such as those of Runge-Kufta, Gear, or Beeman is a result of performance evaluations and of the following considerations.

1. Number of force evaluations per time step. By far the most expensive part of a MD simulation is the force calculation cf. Eq. (40)., which therefore should not be carried out more than once per integration time step At. This rules out algorithms of the Runge-Kutta type.

2. Order ojthe algorithm. It can be shownr69*1271 that the leap-frog, Verlet and Beeman algorithms generate exactly the same trajectory, and are of 3rd-order accuracy in the time step At. The application of higher-order, more accurate algorithms like those of Gear,which involve the use of higher-order derivatives of the function to be integrated, is of no use as long as this function is a non-harmonic, noisy one. For example, in simulations of polar liquids or macromolecular solutions noise due to the cut-off applied to the longrange Coulomb forces prohibits an increase in accuracy beyond 3rd order.

3. However, if the highest-frequency motions in the molecular system are highly harmonic, as in solids, the higher-order function derivative will possess predictive power, which may lead to improved accuracy

by applying higher-order algorithms. Algorithms that are of lower than third order are not efficient in the MD of molecular systems.

This is because molecular potential energy functions V generally have positive second derivatives, requiring third- or higher-order algorithms.

Representation of the algorithm. The much used Verlet algorithm can be obtained from the leap-frog scheme (56)-(57). by eliminating the velocities $v_i(t_n + \Delta t/2)$ and $v_i(t_n - \Delta t/2)$ from Equations (56) and (57) and replacement of t_n by $t_n - \Delta t$ in Equation (57).

$$r_i(t_n + \Delta t) = 2r_i(t_n) - r_i(t_n - \Delta t)$$
$$+ m_i^{-1} F_i(\{r_i(t_n)\})(\Delta t)^2 + O(\Delta t^4) \tag{58}$$

The atomic velocity does not occur explicitly in this algorithm, which makes a coupling of the system to a heat bath by velocity scaling as described in Section 3.4 impossible.

Memory and computational requirements. In some studies the Beeman algorithm Eqs. (59) and (60). is advocated

$$r_i(t_n + \Delta t) = r_i(t_n) + v_i(t_n)\Delta t$$
$$+ m_i^{-1}[4F_i(t_n) - F_i(t_n - \Delta t)](\Delta t)^2/6 \tag{59}$$

$$v_i(t_n + \Delta t) = v_i(t_n) + m_i^{-1}[2F_i(t_n + \Delta t) + 5F_i(t_n)$$
$$- F_i(t_n - \Delta t)]\Delta t/6 \tag{60}$$

Although Equations (59) and (60) are much more complicated than Equation (58), the latter follows directly from the former: replace t_n by $t_n - \Delta t$ in Equations (59) and (60), multiply the latter by Δt and subtract the former from it, and add the resulting equation to Equation (59). Reshuffling of terms then gives Equation (58). Since the Beeman, Verlet and leap-frog algorithms generate identical trajectories, we prefer to use the latter because of its minimal computer memory storage and computational requirements.

Integration Schemes for Stochastic Dynamics

Langevin's equation of motion (41) differs from Newton's equation (54) by the occurrence of a stochastic force $R_i(t)$ and a frictional force $m_i \gamma_i v_i(t)$ Eq. (61)..

$$dv_i(t)/dt = m_i^{-1} F_i(\{r_i(t)\}) + m_i^{-1} R_i(t) - \gamma_i v_i(t) \tag{61}$$

The solution of this equation around = fn is formulated in Equation (62)

$$v_i(t) = v_i(t_n) e^{-\gamma_i(t-t_n)}$$
$$+ m_i^{-1} e^{-\gamma_i(t-t_n)} \int_{t_n}^{t} e^{-\gamma_i(t_n-t')} [F_i(t') + R_i(t')] dt' \tag{62}$$

Since the stochastic properties of $R_i(t')$ are given, the integral over $R_i(r')$ can be obtained directly. The integral over the systematic force $F_i(t')$ is, as in the previous section, obtained by expanding Fi(t') in a Taylor series around t = t, and omitting all terms beyond third order in At in the positions, beyond second order in the velocities, and beyond second order in the forces. The SD equivalent of the leap-frog velocity formula (56) then becomes:

$$v_i(t_n + \Delta t/2) = v_i(t_n - \Delta t/2) e^{-\gamma_i \Delta t}$$
$$+ m_i^{-1} F_i(t_n)[1 - e^{-\gamma_i \Delta t}]/(\gamma_i \Delta t)$$
$$+ V_i(t_n; \Delta t/2) - e^{-\gamma_i \Delta t} V_i(t_n; -\Delta t/2) \tag{63}$$

With

$$V_i(t_n; \Delta t/2) \equiv m_i^{-1} e^{-\gamma_i \Delta t/2} \int_{t_n}^{t_n + \Delta t/2} e^{-\gamma_i(t_n-t')} R_i(t') dt' \tag{64}$$

When yi goes to zero, Equation (63) reduces to Equation (56). The SD equivalent of the leap-frog position formula (57) is obtained by integrating Equation (55) using the velocity expression (62) for the

Eqs. (65), (66)..

$$r_i(t_n + \Delta t) = r_i(t_n)$$
$$+ v(t_n + \Delta t/2) \Delta t [e^{+\gamma_i \Delta t/2} - e^{-\gamma_i \Delta t/2}]/(\gamma_i \Delta t)$$
$$+ X_i(t_n + \Delta t/2; \Delta t/2) - X_i(t_n + \Delta t/2; -\Delta t/2) \tag{65}$$

$$X_i(t_n; \Delta t/2) \equiv (m_i \gamma_i)^{-1} \int_{t_n}^{t_n + \Delta t/2} [1 - e^{-\gamma_i(t_n + \Delta t/2 - t')}] R_i(t') dt' \tag{66}$$

When y_i goes to zero, Equation (65) reduces to Equation

When using the SD leap-frog algorithm (63-66) it must be noted that $V_i(t_n; -\Delta t/2)$ is correlated with $X_i(t_n - \Delta t/2 \ \Delta t/2)$ since they are different integrals of Ri(t) over the time interval $(t_n - \Delta t/2; \ t_n)$ The same observation holds for $X_i(t_n + \Delta t/2; -\Delta t/2)$ and $V_i(t_n; \Delta t/2)$ [cf. Eqs. (65) and (63).. These are different integrals of Ri(t) over the time interval $(t_n, t_n + \Delta t/2)$ This means that these correlated quantities must be sampled in a correlated manner,l' 281 which makes the SD leap-frog algorithm more complicated than the MD one. Integration schemes for the generalized Langevin equation involving time-dependent friction coefficients yi can be found in Refs

Application of Constraints in Molecular Dynamics

In Section 3.4 methods to exactly constrain the temperature and pressure of a molecular system were briefly mentioned. Here, methods to constrain molecular bond lengths or bond angles in dynamic simulations will be discussed. They are used to save computing time. The length of the time step Δt in a MD or SD simulation is limited by highest frequency (vmax) motions occurring in the system Eq. (67)1

$$\Delta t \ll v_{max}^{-1}.$$

(67)

By freezing the generally uninteresting high-frequency internal vibrations, such as bond-length or possibly bond-angle vibrations, v_{max}^{-1}, is increased, which allows for a longer time step Δt. The application of constrained dynamics makes sense physically and computationally when

1. the frequencies of the frozen (constrained) degrees of freedom are (considerably) higher than those of the remaining ones, thereby allowing a (considerable) increase of Δt,

2. the frozen degrees of freedom are only weakly coupled to the remaining ones, i.e. when the molecular motion is not significantly affected by application of constraints,

3. so-called metric tensor effects play a minor role, 4. the algorithm by which the constraints are imposed on the molecular system does not require excessive mathematical or computational effort.

4. the algorithm by which the constraints are imposed on the molecular system does not require excessive mathematical or computational effort.

In molecular simulations typically a factor of 3 in computer with .- time can be saved by application of constraints.

1261 The 1, application of constraints implies that the atoms move on hypersurface in configuration space. While in the full Cartesian configuration space each configuration has equal weight in the partition function, the weighting on the hypersurface will in general depend on its form. The metric tensor defining the hypersurface in terms of Cartesian coordinates will determine these weights. Therefore, when simulating using constraints, metric tensor corrections must be taken into account. Their significance depends on the type of constraint. A discussion of metric tensor effects is given in Refs

Several methods exist for integrating the equations of motion of a molecular system in the presence of constraints. They can be classified as follows:

1. Formulation in terms of generalized coordinates. In this case the constrained degrees of freedom (e.g. bond lengths, bond angles) are treated as fixed parameters, not as degrees of freedom. For example, a macromolecule may be modeled by only torsional angle degrees of freedom, as is often done in static modeling studies. Two cases can be distinguished.

•   When dealing with rigid molecules, center of mass coordinates and Euler angles can be used as degrees of freedom, leading to the Newton-Euler equations of rigid body motion. A variety of algorithms have been proposed for the integration of these equitation.

•   of freedom becomes larger than a few, and when inertial terms are not neglected, it is a tedious task to write down explicitly the appropriate equations of motion in generalized coordinates.

•   ' 363 371 The reason is that the base vectors of the coordinate system become dependent on time, which has a number of unpleasant consequences.

•   '361 So we think that the use of generalized coordinates, such as torsional angles, is highly unpractical for describing the dynamics of macromolecules.

2. Formulation in terms of Cartesian coordinates. Two methods are available for integrating the Cartesian equations of motion of flexible molecules subject to holonomic scleronomous constraints, that is, constraints that are only dependent on atomic coordinates, not on time.

1. Matrix methods

2. 1381 bear their name because they involve the (costly) inversion of a matrix of dimension equal to the number of constraints. Thus, these methods are not well suited for application to macromolecular systems.

3. Iterative methods, like the so-called SHAKE method," 381 are especially appropriate for macromolecules, since they treat the constraints in an iterative way.

Algorithms for Constraint Dynamics in Cartesian Coordinates

Molecular constraints have the form (68) for the case of

$$\sigma_k(r_1, \ldots, r_N) = 0 \qquad k = 1, \ldots N_c$$

(68)

N_c constraints in a molecule consisting of N atoms. Bondlength and bond-angle constraints can be put in the form of distance constraints between atoms k_1 and k_2 as in (69) where the constraint distance is given by d_{k1k2}1

$$r_{k_1k_2}^2 - d_{k_1k_2}^2 = 0 \tag{69}$$

When applying Constraints in MD or SD, the 3N equations of motion (39) or (41) have to be integrated while satisfying the N, constraints. This can be accomplished by applying Lagrange's method of undetermined multipliers. A zero term (68) is added to the potential energy function V in Equation (40), which yields (70) as the equation of motion.

$$m_i\, d^2 r_i(t)/dt^2 = -\frac{\partial}{\partial r_i}\left\{ V(\{r_i(t)\}) + \sum_{k=1}^{N_c} \lambda_k(t)\, \sigma_k(\{r_i(t)\}) \right\} \tag{70}$$

The time-dependent multipliers $\lambda_k(t)$ are determined such that the constraints σ_k are satisfied. The physical interpretation of Equation (70) becomes clear by rewriting it in terms of forces Eq. (71).:

$$m_i\, d^2 r_i(t)/dt^2 = F_i(t) + G_i(t) \tag{71}$$

The total unconstrained force $F_i(t)$ derived from the potential energy function is the first term in Equation (70), while the constraint force $G_i(t)$), which compensates the components of $F_i(t)$ that act along the directions of the constraints, is the second term.

The leap-frog scheme (56)-(57). for integration of Equation (71) becomes Equations (72) and (73).

$$v_i(t_n + \Delta t/2) = v_i(t_n - \Delta t/2) + m_i^{-1}\{F_i(t_n) + G_i(t_n)\}\Delta t \tag{72}$$

$$r_i(t_n + \Delta t) = r_i(t_n) + v_i(t_n + \Delta t/2)\Delta t \tag{73}$$

Separating contributions from $F_i(t_n)$ and $G_i(t_n)$ we find

$$r_i(t_n + \Delta t) = r_i' + \delta r_i \tag{74}$$

with

$$\delta r_i = m_i^{-1} G_i(t_n)(\Delta t)^2 \tag{75}$$

where r_i' are the positions after a MD or SD step disregarding all constraints, and 6vi are the positional corrections to be made as a result of the constraints. Using the definition of G_i we find

$$\delta r_i = - m_i^{-1}(\Delta t)^2 \sum_{k=1}^{N_c} \lambda_k(t_n) \frac{\partial}{\partial r_i} \sigma_k(\{r_i(t_n)\})$$

(76)

or when using the explicit form (69) for σ_k,

$$\delta r_i = - 2m_i^{-1}(\Delta t)^2 \sum_{\substack{k=1 \\ (k_1, k_2) = (i, j)}}^{N_c} \lambda_k(t_n) r_{ij}(t_n)$$

(77)

where the summation extends only over constraints involving atom i. This implies that corrections due to the distance constraint between atoms i and j must be applied in the directions of the vector rij. Corrections to r_i and r_i are in opposite directions and weighted by the inverse mass of atoms i and j, as is illustrated in Figure 12. Since the positions $r_i(t_n + \Delta t)$ must satisfy the constraints (68, 69) using (74) we have for each constraint:

$$[r'_{k_1} + \delta r_{k_1} - r'_{k_2} - \delta r_{k_2}]^2 = d^2_{k_1 k_2}$$

(78)

These form a set of N_c quadratic equations from which the N_c Lagrangian multipliers λ_k can be determined. After linearization of Equation (78) by neglecting the terms quadratic in λ_k, a set of linear equations is obtained. The above mentioned matrix methods solve these by a matrix inversion.

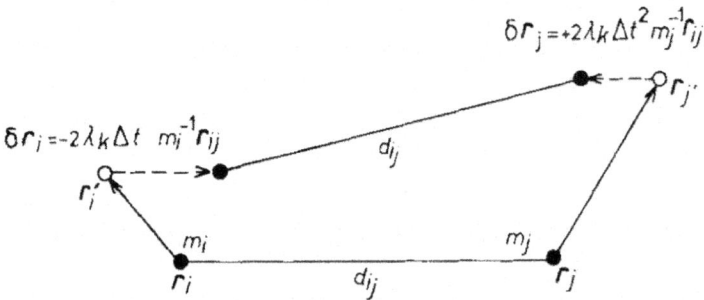

Figure 12: Application of distance constraints. Atomic coordinate resetting using Lagrangian multipliers λ_k

1381 The iterative methods solve them by treating all constraints in succession, and iterating this procedure until all constraints are satisfied within a specific geometric tolerance. The so-called SHAKE method, which is much used, is of the iterative type." 381 Its application in various algorithms is denoted by Equation (79). This means that the positions

$$\text{SHAKE } (r_i(t_n), r'_i(t_n + \Delta t), r_i(t_n + \Delta t))$$

(79)

$r'_i(t_n + \Delta t)$resulting from a non-constraint time step will be reset to give the constrained positions $r_i(t_n + \Delta t)$. When necessary the constraint forces can be determined from Equation (80) and the constrained velocities from Equation (81)

$$G_i(t_n) = m_i[r_i(t_n + \Delta t) - r'_i(t_n + \Delta t)]/(\Delta t)^2 \tag{80}$$

$$v_i(t_n + \Delta t/2) = [r_i(t_n + \Delta t) - r_i(t_n)]/\Delta t \tag{81}$$

Effect of the Application of Constraints

The effect of constraining bond lengths and bond angles in molecular systems has been evaluated. turns out that the application of bond-length constraints saves about a factor of 2 in computing effort when bonds to hydrogen atoms are constrained and about a factor of 3 when all covalent bonds are constrained. No evidence was found for a distortion of the physical properties by the rigidity of the bonds." 391 When the bond angles undergo limited variation, which is generally true for molecular systems, metric tensor corrections play an insignificant role when bond-length constraints are applied.

The use of bond-angle constraints is not allowed, since it considerably affects the molecular motion. Macromolecular flexibility and entropy are halved, and the number of torsional angle transitions is dramatically reduced.

139. Moreover, it has been shown that in case of application of bond angle constraints metric tensor corrections, which are nearly impossible to calculate for all but the smallest flexible molecules, are of significant size and may not be ignored.

This means that MD or SD of molecular systems should not be performed in torsional space while treating bond lengths and angles as fixed quantities.

Disadvantages of the Application of Constraints

The advantage of the application of bond-length constraints is clear: at the expense of about 10% extra computing time, longer, time steps Δcan be taken. When simulating macromolecular systems without constraints. a value of $\Delta t = 0.5$ fs is appropriate, whereas with bonds to hydrogen atoms constrained $\Delta t = 1.0$ fs, and with all bonds constrained $\Delta f = 2.0$ fs is appropriate. So, a factor of 2 to 4 in computing effort is saved.

Yet, the application of constraints also has its disadvantages.

1.   1Convergence problems for large planar groups. In practice, procedures like the SHAKE method sometimes fail to converge to a molecular

configuration satisfying all constraints. This is often due to the fact that the constraint forces act along the bond directions in the previous MD step (Fig. 12). For a planar group of atoms the constraint forces will act along vectors in the plane. So, when the other forces, e.g. due to charge repulsion, act orthogonally to the plane. it is nearly impossible for the constraint forces to counteract these, since they act orthogonally to each other. A solution to this problem has been proposed in Ref.

2. 140., where virtual atoms lying outside the plane of the planar group are used to obtain constraint forces with sizeable components orthogonal to the plane of the group of real atoms. The implementation of this virtual atom technique would require definition of virtual atoms for each possible type of planar group. This makes this technique not very attractive for application in macromolecules.

3. Free energy of creation or annihilation of atoms. When free energy differences are computed using the coupling parameter approach, the application of constraints is a complicating factor. The Hamiltonian, or the potential energy function, is made a function of a coupling parameter i, which is smoothly changed in the course of a simulation in order to obtain the work done by the system over a reversible change of i.. The problem now is that a constrained bond length cannot be removed from the system as a smooth function of 1. When computing the free energy of breaking a bond one faces the fact that the removal of a constraint is a discontinuous process. Even when only the length of a bond is changed (not removed) as a function of L, computation of the free energy change due to the work done by the constraint forces is not simple, as is shown in Ref. 141..

4. Computational botrleneck for parallelization of' algorithms. The constraint forces are generally computed after the other forces have been calculated and an unconstrained integration time step is performed. This makes the application of constraints a computational bottleneck : all other computations have to wait for its termination. Moreover, the most efficient methods for application of constraints, such as the SHAKE method, are of an iterative nature which makes them unsuitable for parallelization on a computer.

5. Physical model of frozen bonds. Although no significant effects due to freezing of bonds have been observed for molecular systems in equilibrium, it remains to be seen whether in non-equilibrium situations the finite flexibility of bonds may play a role in the dynamics, e.g. as an energy reservoir.

Multiple Time Step Algorithms

An alternative to the application of bond length constraints is the use of a multiple time step (MTS) integration algorithm. The length of the integration time step is limited by the oscillation or relaxation time of the forces see Eq. (67).. In a molecular system three frequency ranges can be distinguished, as shown in Table 5: high-frequency bondstretching forces F^{hf}, low-frequency long-range Coulomb forces F^{lf} and the remaining intermediate frequency forces F^{if}. The contribution of the different forces to the atomic trajectories may be integrated using different time steps. An example is the twin range method discussed in Section 3.1.3.4, in which the long-range Coulomb force is kept constant during k time steps At, where k lies in the range 5- 100

Table 5: Various relaxation times in macromolecular systems, and force components to be used in multiple time step algorithms

Type of force	Approximate oscillation or relaxation time (fs)	Force		
		Application at step $n' = 0,1,2,\ldots$	Update at step	Size of force to be applied
1. high frequency: F^{hf} covalent bond stretching	10	$t_n + n'\Delta t'$	$t_n + n'\Delta t'$	F_i^{hf}
2. intermediate frequency: F^{if} bond-angle bending dihedral angle torsion van der Waals shortrange Coulomb	40	$t_n + n'm\Delta t'$	$t_n + n'm\Delta t'$	mF_i^{if}
3. low frequency: F^{lf} long-range Coulomb	1000	$t_n + n'm\Delta t'$	$t_n + n'mk\Delta t'$	mF_i^{lf}

A multiple time step scheme could also be applied to the bond-stretching forces.The conventional MD or SD time step Δt is subdivided into m (m = odd) smaller substeps $\Delta t'$, so $\Delta t = m\Delta t'$. The covalent bond forces F^{hf} are evaluated at each time step $t_n + n'\Delta t'$, with $n' = 0,1,2,\ldots$ When n' a multiple of m, the other forces $F^{if} + F^{lf}$ are applied, but multiplied by a factor m in order to compensate for their omission at the substeps. Application of this MTS algorithm to macromolecules shows that for in = 3 or 5 the integration is as accurately performed as when bond-length constraints are used. Since the bond frequencies are about four times the other ones, a value of m = 4 would be expected to be large enough to ensure proper integration of the bond forces.

We note that the physical quantities one is interested in should only be evaluated at the conventional time steps Δt, not at the sub steps $\Delta t'$. Since the bond stretching forces can be rapidly computed, application of this MTS algorithm is as efficient as the use of the SHAKE method, but it avoids the disadvantages of the latter method which were discussed in the previous section.

Searching Neighbors

The bulk (approximately 9Oo/o) of the computer time required by a MD or SD simulation is used for calculating the nonbonded interactions, that is, for finding the nearest neighbor atoms and subsequently evaluating the van der Waals and Coulomb interaction terms for the atom pairs obtained. Various schemes for performing this task as efficiently as possible have been proposed.

1. I Neighbor list techniques. Once the neighbors have been found, either by scanning of all possible atom pairs (operation proportional to N^2), or by using grid-search techniques (proportional to N), the pairs are stored in a neighbor list which is only updated every so many steps, typically every 5 to 100 steps. At each time step the neighbor list is used to calculate the interactions (operation proportional to N).

2. Grid search techniques. The computational box is filled with a grid or mesh, and for each grid cell it is determined which atoms lie in it. This operation is proportional to N. The nearest neighbors of an atom can easily be found in the grid cells surrounding the grid cell containing the atom.

An evaluation of the different schemes can be found in Ref. For small systems $(N \lesssim 1000)$ the use of neighbor list techniques will reduce finding the neighbors to a small part of the calculation. For large systems $(N \lesssim 1000)$ the application of grid search techniques will become advantageous.

Initial Conditions, Equilibration, and Analysis of Simulations

A simulation starts with initial atomic positions and velocities. The results should be independent of these. The initial configuration for a molecular system can be obtained from different sources: X-ray structure, model building, distance geometry calculation, random search techniques, etc. The initial velocities are either taken from a Maxwellian distribution or chosen to be zero, in which case strain in the molecule that is converted into kinetic energy may generate non-zero velocities.

The equilibration period that is required will depend on the relaxation time of the property one is interested in. Some properties, such as the kinetic energy, require short (picoseconds) equilibration times, whereas others, such as dielectric properties, may require longer times of the order of tens of picoseconds. During a simulation a number of quantities, such as the potential and kinetic energy, or the diffusion from the initial structure, are generally monitored to obtain a picture of the stability of the simulation.

The results are generally analyzed by taking time averages or averages over simulations with different initial conditions of the quantities of interest. Fluctuations and correlation functions may be calculated to analyze the mobility and dynamic behavior of the system.

APPLICATION OF SIMULATIONS

Understanding in Terms of Atomic Properties

The obvious utility of computer simulation is the possibility of analyzing molecular processes at the atomic level. Many examples can be found in the literature here, we briefly mention two recent examples, the atomic interpretation of biochemical data on repressor-DNA operator binding, and the atomic interpretation of biophysical measurements on membrane properties.

Interpretation of Biochemical Data on Repressor-Operator Binding by MD Computer Simulation

In Ref. a 125-ps MD simulation of a Lac repressor headpiece (51 amino acid residues) complexed with its DNA operator (14 base pairs) in aqueous solution was reported. The starting structure for the simulation was obtained from model building and energy minimization based on 169 proton-proton distances for the headpiece and 24 headpieceDNA proton-proton distances which were available from 2D-NMR measurement. When analyzing the repressoroperator contacts, no evidence was found that a so-called "direct readout" mechanism'1 for recognition is based on direct repressor side-chain-base hydrogen bonds, which is the common view on recognition. The simulation suggested that direct readout occurs rather through non-polar contacts and water-mediated hydrogen bonds. The repressor-operator contacts as observed in the simulation are compatible with the available biochemical data on base-pair or amino acid residue substitution, but do not support the usual interpretation in terms of side-chain-base hydrogen bonds. This illustrates the usefulness of computer simulation studies to obtain an interpretation of biochemical data at the atomic level.

Interpretation of Biophysical Data on Membrane Properties by MD Computer Simulation

In Ref. a 180-ps MD simulation of a bilayered system of 52 sodium ions, 52 decanoate, 76 decanol and 526 water molecules was reported. A detailed analysis of the lipid-water interface was made. The charged (decanoate) head groups lie well within the water layer, but the alcoholic groups are situated more on the lipid side of the interface, which appears to be very diffuse and extends over almost 10 A. Unexpectedly, the overlap of sodium ion and carboxylic acid distributions suggests a charge compensation rather than an electric double layer as pictured in most textbooks. Water molecule orientation is such that the remaining ion charge distribution is compensated.

Lateral diffusion constants of the lipid molecules as measured in the simulation $(3 \times 10^{-6} \, \text{cm}^2 \, \text{s}^{-1})$ compare well with experiment $(2 \times 10^{-6} \, \text{cm}^2 \, \text{s}^{-1})$ as measured using nitroxide spin labels. Analysis shows that the hydrodynamic interaction of the head groups with the aqueous layer determines the diffusion constant of the lipid molecules, rather than interactions within the lipid layer. This illustrates the usefulness of computer simulation studies to obtain an interpretation of biophysical data at the atomic level.

Determination of Spatial Molecular Structure on the Basis of 2D-NMR, X-Ray or Neutron Diffraction Data

During the last few years computer simulation has become a standard tool in the determination of spatial molecular structure on the basis of X-ray or neutron diffraction or 2D-NMR data. The goal of structure determination based on experimental NMR or diffraction data is to find a molecular structure that

1. satisfies the experimental data, such as a set of atomatom distance constraints $\{r_{ij}^0\}$ or torsional angle constraints $\{\varphi_{ij}^0\}$ in the case of NMR, or a set of observed structure factor amplitudes $F_{obs}(hkl)$ and, if available, phases $\alpha(hkl)$ in the case of diffraction data, and

2. has a low energy in terms of a molecular potential energy function $V_{phys}(\{r_i\}$ Eq. (1).

3. In order to optimize a structure simultaneously with respect to both criteria, the experimental information is cast into the form of a penalty function or restraining potential V_{restr} ,, the value of which increases the more an actual structure violates the experimental data. The most simple choice of a function taking into account maximum values for the interatomic distances $\{r_{ij}^{ub}\}$ would be Equation (82), where the force constant is denoted by K_{dr} .

$$V_{restr} = V_{dr} \equiv \tfrac{1}{2} K_{dr} \sum_{\text{all constraints } (ij)} [\max (0, r_{ij} - r_{ij}^{ub})]^2$$
(82)

The corresponding function which restrains the calculated structure factor amplitudes F,,,(hkl) to the observed ones is defined in (83).

$$V_{restr} = V_{sf} \equiv \tfrac{1}{2} K_{sf} \sum_{hkl} [F_{calc}(hkl) - F_{obs}(hkl)]^2$$
(84)

The optimization problem is to find a molecular structure for which the energy function (84) attains the global

$$V_{opt} = V_{phys}(\{r_i\}) + V_{restr}(\{r_i\})$$
(85)

minimum. As discussed in Section 3.2, MD simulation is a very powerful method to search configuration space for low energy configurations, due to its ability to surmount energy barriers of the order of $k_B T$ per degree of freedom. Therefore, the application of MD computer simulation in spatial structure refinement of 2D-NMR or X-ray diffraction data has become widespread in recent years.

Refinement of Structures Based on NMR Data

After introduction of the method of MD refinement it has been applied to a variety of molecules using different refinement protocol Th e penalty function V_{restr} may be chosen in different ways. The relative weight of the restraining term in Equation (84), that is the value of K_{dr} may be chosen so large as to reduce the distance violations, at the expense of increasing the intramolecular energy V,,, . Too large a K_{dr} value will generally lead to a strained (unphysical) molecular structure. The searching of configuration space may be performed at high temperature in order to more easily cross energy barriers.

The standard procedure is to start with a number of distance geometry (DG) type calculations to generate a collection of starting structures, which are subsequently refined by MD simulation. In Ref. it was shown that the best DG structure in terms of distance violations is generally not the best structure after MD refinement. This is due to the crude energy function used in DG which has to be cast in the form of distances, and so favors geometries which are unfavored by more sophisticated energy functions used in simulations.

When the NMR data contain contributions from different molecular conformations it will be impossible to find one conformation satisfying the experimental data." 5s1 This observation led to the introduction of time-dependent constraints" s61 which do not force the molecule to satisfy the distance constraints at each time point of the simulation, but only forces the distance constraints to be satisfied on average. This forms a better representation of the measured data.

Refinement of Structure Based on Crystailographic X-Ray or Neutron Diyfraction Data

Since its introduction the method of MD refinement of crystallographic data" 571 has been applied mainly in protein crystallography. Again, different penalty functions (83) and refinement protocols are used. Generally speaking MD refinement saves a few months work over conventional refinement when applied to proteins. A survey of recent developments is given in Ref. As in the NMR case, the use of time-dependent structure factor restraints will be a much better representation of the experimental information, which is an average over time and over molecules. Preliminary results (P. Gros, private communication) indicate that lower R values and better searching of conformational space are obtained.

Prediction of Structural Changes by MD Simulation

No reliable approaches are as yet available for the prediction of de novo (macro) molecular structures, and there is no real alternative to experimental structure determination. When structural information on homologous molecules is available, a molecular structure may be predicted starting from the homologous molecules and subsequently changing it to the required composition, either by model building on a graphic device or by more automatic simulation techniques. When additional experimental information from Xray diffraction or NMR is available the chance of success is greatly increased. An example is the structure determination of the 279-residue protein thermitase, starting from the known structure of subtilisin (275 amino acid) These proteins show only 47 % homology; residues had to be changed, deleted and inserted when changing subtilisin into thermitase. Subsequent MD refinement using crystallographic X-ray data on thermitase made numerous atoms move over more than 4 A. Some parts showed a structural change of more than 8 A.

If only one or a few amino acid side chains in a protein are changed (mutated), the structural change may be predicted by MD simulation without the extra help of X-ray data. In Figure 13 the conformation of the Met-222 + Phe mutant of subtilisin resulting from a MD simulation in which Met-222 was gradually changed into Phe-222 is compared with the conformation as obtained from an X-ray diffraction determination of this subtilisin mutant." 601 The predicted conformation is correct. However, this is not always true; especially when hydrogen bond networks, involving bound water molecules, have to be rearranged, the simulation period may be too short compared to the hydrogen bond network relaxation time.

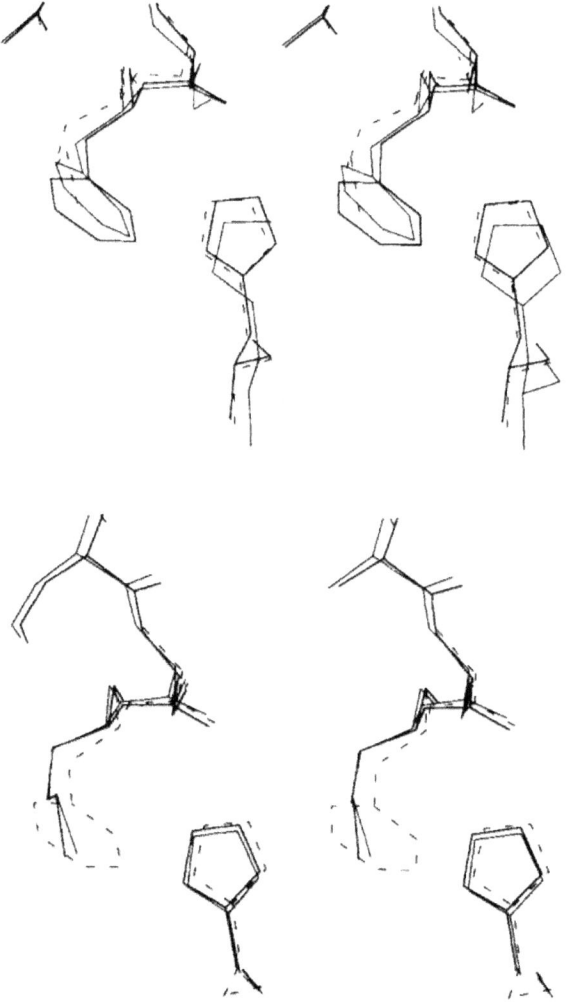

Figure 13: Stereoview of the hydrophobic binding pocket of subtilisin. Upper part: forward 222 Met -t Phe mutation; dashed line: 222 Met (native) starting structure; thick solid line: 222 Phe (mutant) MD predicted structure; thin solid line: 222 Phe (mutant) X-ray structure. Lower part: backwards 222 Phe + Met mutation; dashed line 222 Phe (MD) starting structure; thick solid line: 222 Met MD predicted structure (from 222 Phe mutant); thin solid line 222 Met (native) MD structure (from X-ray).

It is also possible by MD simulation to determine the conformation of a molecule as a function of the type of environment, viz. crystalline, nonpolar solution, aqueous solution, etc. Studies of this type have been performed for different molecules.161 - 1641 For the immuno suppressive drug cyclosporin A (CPA) the conformational differences found between its crystal conformation and its conformation in a chloroform solution were exactly reproduced by two MD simulations in the corresponding environments (Figure 16).

This application illustrates the power of MD simulation when studying conformational properties of flexible molecules or mutants. However, when large conformational changes are to be expected the length (typically 10'-lo2 ps) of a MD simulation may be too short to bring these about in the limited time available to surmount possible energy barriers.

Prediction of Free Energy Changes by MD Simulation

From an MD trajectory the statistical equilibrium averages can be obtained for any desired property of the molecular system for which a value can be computed at each point of the trajectory. Examples of such properties are the potential and kinetic energy of relevant parts of the system, structural properties and fluctuations, electric fields, diffusion constants, etc. A number of thermodynamic properties can be derived from such averages. However, two important thermodynamic quantities, the entropy and the (Gibbs or Helmholtz) free energy, generally cannot be derived from a statistical average. They are global properties that depend on the extent of phase (or configuration) space accessible to the molecular system. Therefore, computation of the absolute free energy of a molecular system is virtually impossible. Yet, the most important chemical quantities like binding constants, association and dissociation constants, solubilities, adsorption coefficients, chemical potentials, etc., are directly related to the free energy. Over the past decade statistical mechanical procedures have evolved by means of which relative free energy differences may be obtained. For recent reviews of methodology and applications see Refs. The most powerful method is the so-called thermodynamic cycle integration technique. The free energy difference between two states A and B of a system is determined from a MD simulation in which the potential energy function $V(\{r_i\})$

Eq. (I). is slowly changed such that the system slowly converts from state A into state B. In principle the free enthalpy difference $\Delta G_{BA} = G(B) - G(A)$ is determined as the work necessary to change the system from state A to state B via a reversible path. An example is given in Figure 14, where the free enthalpy of changing one water molecule (state A) in a cube with 216 water molecules into a methanol molecule (state B) is illustrated. The change is carried out over a period of 20 ps and yields $\Delta G_{BA} = 6.1 \text{ kJ mol}^{-1}$, as compared with an experimental value of 5.2 kJ mol^{-1}.

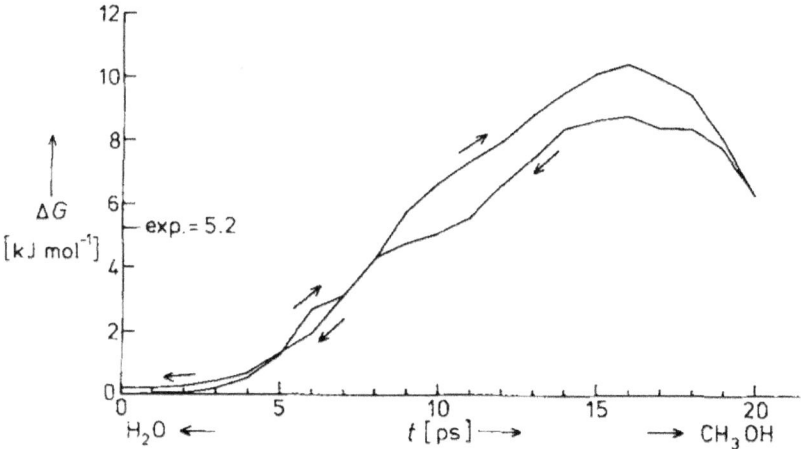

Figure 14: Free enthalpy change AC on changing H,O into CH,OH and back again.

The basis on which the thermodynamic cycle approach rests is the fact that the free enthalpy G is a thermodynamic state function. This means that as long as the system is changed in a reversible way the change in free enthalpy ΔG will be independent of the path. Therefore, along a closed path or cycle one has $\Delta G = 0$. This result implies that there are two possibilities of obtaining AG for a specific process; one may calculate it directly using the technique sketched above along a path corresponding to the process, or one may design a cycle of which the specific process is only a part, and calculate the AG of the remaining part of the cycle. The power of this thermodynamic cycle technique lies in the fact that on the computer also non-chemical processes such as the conversion of one type of atom into another type (H into CH,, see Fig. 14) may be performed.

The method is outlined in Figure 15 for the process of binding of two different inhibitors, trimethoprim (TMP) and its triethyl analogue (TEP), to the enzyme dihydrofolate reductase (DHFR) from chicken liver

'66. 1671 in the presence of coenzyme (NADPH) and water. The appropriate thermodynamic cycle is given in Figure 15. The relative binding constant of the two inhibitors equals K,IK, = $\exp[-(\Delta G_2 - \Delta G_1)/RT]$.

However, simulation of processes 1 and 2 is virtually impossible, since it would involve the reversible removal of many solvent molecules from the active site of theenzyme to be substituted by the inhibitor. Since the processes in Figure 15 form a cycle, one has $\Delta G_2 - \Delta G_1$ = AG4 - AG3, and the processes 3 and 4 can easily be simulated since they involve the change of three oxygen atoms (TMP) into three CH, groups (TEP). MD simulation of process 3 yields

$\Delta G_3 = -61 \pm 0.2 \text{ kJ mol}^{-1}$ and of process 4 yields $\Delta G_4 = -65 \pm 10 \text{ kJ mol}^{-1}$. So, the computed difference in free enthalpy of binding between TEP and TMP is $+4 \pm 10 \text{ kJ mol}^{-1}$ to be compared with an experimental value of $+7 \text{ kJ mol}^{-1}$.

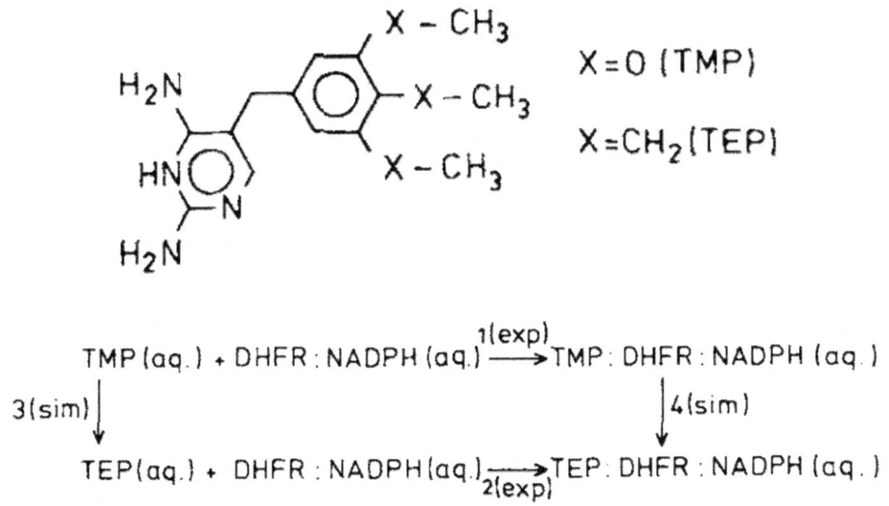

Figure 15: Thermodynamic cycle for computation of the relative free enthalpy of binding of two inhibitors, trimethoprim (TMP) and its triethyl analogue (TEP), to the enzyme dihydrofolate reductase (DHFR) in the presence of coenzyme (NADPH) in aqueous solution. Complexation is denoted by the symbol ":".

Although this result seems reasonable, the computed value may only be considered as an order-of-magnitude estimate. This is due to various assumptions and approximations that underlie this type of free energy calculation. Since these have been discussed elsewhere, the most important ones will only be summarized here.

Adequate sampling, or the relaxation time oj the environment. The change from state A to state B has to be carried out in a reversible way, which means that the time period of the change must be much longer than the relaxation time of the environment which has to adapt to the change. The rotational correlation time of a water molecule is about 2 ps and the dielectric relaxation time of water is about 8ps. This means that a MD simulation of the change from A to B over 20 ps or more is long enough to obtain reasonably accurate ΔG values (see Fig. 14).

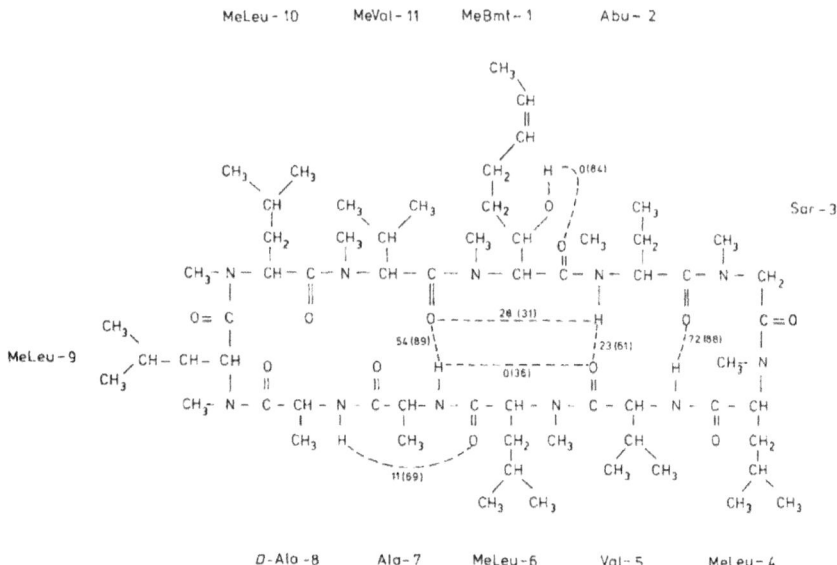

Figure 16. Hydrogen bonds (dashed lines) occurring in two MD simulations of cyclosporin A. Percentage of occurrence is given, both for a MD simulation in aqueous solution and (in parentheses) for a MD simulation in vacuo.

However, in process 4 (Fig. 15) the relaxation time of the (protein) environment of the inhibitor is much longer, which results in a large uncertainty of 10 kJ mol⁻¹.

Effect of long-range Coulomb interactions. Free energy computation involving the creation or annihilation of (full) atomic charges is very dependent on a proper treatment of the Coulomb interaction, which has an r-' distance dependence. For example, when creating an ion in aqueous solution, the contribution of the hydration shell between 9 8, and 12 8, to the free energy of solvation is about 40 kJ mol-','4'1 which means that the cut-off radius must be very large to obtain accurate AG values. When only dipolar changes play a role, as in the case of Figures 14 and 15, the interaction has an F3 distance dependence, which considerably reduces the need to use a large cut-off radius.

Sensitivity to force field parameters. Not unexpectedly, the free energy of a molecular system is rather sensitive to the interaction function (1) that is used in the simulation. For example, in the relatively simple case of the free enthalpy of solvation of methanol in water (see Fig. 14), application of force field parameters of established force fields yields values ranging from 0 to 14 kJ mol⁻¹, a spread of $2-3\,kT$ around $G_{exp} = 5.2$ kJ mol⁻¹.

The techniques for computation of (relative) free energy are in many applications not yet sufficiently accurate. However, they are still being improved and are in principle widely applicable in the study of molecular systems.

Static versus Dynamic Representation of a Molecular System

The common approach to modeling a molecular system on a computer is a static one. For example, quantum calculations yield an equilibrium charge distribution, molecular mechanics calculations one or a few minimum energy conformations of a molecule; on a graphics device, molecules are studied in terms of fixed conformations. However, 9 molecular system at room temperature is by no means of static character. It should be described in terms of a multidimensional distribution function of all atomic coordinates and its development in time. An example of a dynamic equilibrium between different hydrogen bond patterns is given in Table 6 for crystalline cyclodextrin. The 061 atom is in volved in different, mutually exclusive hydrogen bonds. The ones that do exist more than about 20% of the MD simulation are also observed, as expected, in the neutron diffraction experiment.

Table 6: Dynamic equilibrium between various hydrogen bond patterns involving the 061 atom in crystalline rr-cyclodextrin

Donor [a]	Acceptor	%	MD simulation distance [Å] D$\cdots$A	angle [°] D-H$\cdots$A	Neutron diffraction distance [Å] D$\cdots$A	angle [°] D-H$\cdots$A
	O 63	79	2.79	160	2.81	177
	O 65	2	3.04	158		
O 61-H 61	O 56	1	2.99	158		
	O W3	6	3.04	146		
	O WA	+ 2	2.08	156		
		90				
O 21-H 21		14	3.06	156		
O 63-H 63		3	2.84	161		
O WA-H WA1		36	2.97	156	2.97	173
O WA-H WA2	O 61	5	3.01	154		
O W3-H W31		8	3.07	153	3.33	107
O W3-H W32		11	3.04	153		
O WB-H WB1		+ 2	2.94	156		
		79				

[a] Abbreviations. D = donor, A = acceptor, W = water atoms [29].

1.   It is also observed that the lower the occurrence of a hydrogen bond is in the simulation, the more deformed its geometry is when described in terms of one static hydrogen bond, as is the case for the neutron diffraction data (see Table 6).

The Role of a Solvent in Molecular Simulations

Many (macro)molecular modeling studies involve an isolated molecule without any solvent. This means that solvent effects are completely ignored. The molecular surface will be distorted by the vacuum boundary condition, and certainly no meaningful free energy estimates can be obtained. When a molecular complex is formed, solvent molecules may play a bridging role, as is observed in the repressor-operator complexes.

26. 1691 Even intramolecular hydrogen bonding is affected by the competitive presence of water molecules surrounding a solute in aqueous solution. An example is given in Figure 16, where the occurrence of intramolecular hydrogen bonds is given for cyclosporin A, on the one hand simulated in vacuo and on the other in aqueous solution. It is clear that omission of the solvent results in too high a percentage of intramolecular hydrogen bonds. Proper treatment of solvent effects is a necessary condition for a reliable simulation of molecular properties.

Other Applications of Computer Simulation in Chemistry and Physics

The field of computer simulation of molecular systems has grown so rapidly that a review of all possible applications is hardly feasible. In this section we have chosen examples, mostly of our own work, to illustrate specific aspects of simulation studies. Many other applications may be found in a number of reviews and monographs on computer simulation in physics and chemistry. We would like to mention a few recent studies which may lead the reader to areas of application that were not mentioned here. Computer simulation has been applied to the study of electrolytes, ionic conductors, ionic crystals, ionic salts, and, inter alia, to processes like adsorption sputtering, and melting.

FUTURE DEVELOPMENTS IN COMPUTER SIMULATION

Quantum Simulations

Classical computer simulation implies a number of restrictions, as mentioned in Section 2.1.3, which are due to the assumption that quantum effects play a minor role. A proper description of low-temperature or light-atom (hydrogen) motion or chemical reactions requires a quantum mechanical treatment. Various quantum methods for application in the area of simulation have been developed and recent years have shown an increasing activity in the area of quantum simulations. Here, we briefly mention a few methods. Quantum effects can be incorporated in simulations in different ways.

1. Quantum corrections to the results of classical simulations can be made. Expansion of the partition function in powers of h, Planck's constant, leads to correction formulas for thermodynamic quantities like the free energy" or structural quantities like the radial distribution function $g(r)$. Berens et al. propose a correction formula based on a harmonic approximation of the atomic motion. Also for time-dependent equilibrium quantities, quantum correction formulas are

2. Quantum mechanical treatment of a few degrees of freedom in an otherwise classical simulation can be implemented in different ways.

• The path-integral sirnulation -yields a quantum mechanical equilibrium distribution. Its name originates from its derivation by a discretization of the path-integral form of the density matrix. A recent application is the treatment of electron transfer reactions.

• The Gaussian wave packet method"""-Is7. constituted an early attempt to solve the time-dependent Schrodinger equation by computer simulation. A problem inherent in this method is that of arriving at an adequate description of the interaction between a wave packet and its classical environment.

• In the density functional dynamic method an energy functional, as used in density-functional theory, is added to the Lagrangian of the molecular system, and the Lagrangian equations of motion are subsequently integrated using MD techniques. Static and dynamic properties of crystalline silicon were obtained in terms of a self-consistent pseudo potential.

• In the adiabatic quantum molecular dynamics method the time-dependent Schrodinger equation for the excess electrons and Newton's equation of motion for the nuclei plus core electrons are integrated in the Born-Oppenheimer approximation. Electronic states and dynamic properties of dilute liquid K-KCl solutions were studied.

The potential of the different methods has not yet been fully explored, which means that a best methodology has not as yet crystallized. The field is in a state of rapid development and holds much promise for a proper dynamic treatment of quantum degrees of freedom in the future.

System Size and Time Scale of Simulations

The number of particles in a computer simulation typically lies in the range of 102-104, although simulations involving more than lo4 atoms are nowadays Relatively complex systems like a protein embedded in a membrane will also contain about 10^5 atoms. In practical applications limited system size is much less a concern than the finite time scale of a computer simulation. Due to large energy barriers in the potential energy surface, it may take a molecular system a very long time to cross these barriers and so to sample configuration space efficiently. Typical simulation periods are 10^1-10^2 psec, which is much too short for a proper description of properties which show a much longer relaxation time. Therefore, possibilities to lengthen the time scale of MD simulations are continuously investigated. Here we mention a few.

1. Freezing of degrees of freedom. The basic problem of removing degrees of freedom is that of defining an appropriate interaction function for the remaining degrees of freedom. Moreover, the degrees of freedom removed should play a minor role in the processes one is interested in. Application of bond-length constraints and the twin range method to handle long-range forces fall in this class. An alternative to freezing of degrees of freedom is the use of multiple time step integration methods.

2. Stochasti$cation of degrees of freedom. Explicit treatment of degrees of freedom may be replaced by the application of a mean force plus a stochastic force which represent the average effect of the removed or stochastified degrees of freedom on the remaining explicitly treated ones. An example is the use of the Langevin equation to model the solvent or solid-state environment of a molecule.

3. Scaling of system parameters. Assume that we are interested in a quantity Q which has a relaxation time z that is longer than the time period that could be covered by a MD simulation. It may be possible to identify one or more system parameters p for which z is a (rapidly) changing function $\tau(p)$ of p can be changed to p_{short} such that $\tau(p_{short})$ becomes shorter than the length of a MD simulation, which can then be carried out to obtain $Q(p_{short})$ Extrapolation of $Q(p)$ for p from p_{short} to p_{phys} will yield an estimate $Q(p_{phys})$

The risk associated with the application of this technique is that by scaling of system parameters, the physical processes are changed such that a process at p_{short} may have nothing to do with one at p_{phys}. For example, when studying the relaxation of vibrational energy of an HCl molecule in an Ar lattice, the relaxation time may be shortened from the ps time scale to the ns time scale by reducing the HCl bond-stretching force constant. This will, however, certainly influence the balance between various relaxation mechanisms that are feasible for this process.

1. Activated barrier crossing. In the case of an activated process, methods exist to avoid a full simulation of the rare event of barrier crossing. The procedure consists of three steps: 1) The location of the barrier must be determined. 2) The likelihood that the system will be at the top of the barrier is computed using umbrella sampling techniques. 3) The transition probability is computed by running MD trajectories from the top of the barrier. This technique has been applied to simulate ring flips in Proteins.

2. Mass tensor dynamics. In the classical partition function the integration over the momenta of the particles can be carried out separately from that over the coordinates, when no constraints are applied. The atomic masses do not appear in the configurational integral, which means that the equilibrium properties of a system will be independent of the masses in the system. The technique of mass tensor dynamics exploits this freedom by choosing the atomic masses such that the high-frequency motions of the molecular system are slowed down, which allows a longer simulation time step to be taken.

None of the above-mentioned methods for lengthening the time scale of a simulation is really satisfactory. Most of the progress in this respect will probably result from the ever increasing power of computers. Yet, a combination of the different techniques mentioned above may yield a considerable reduction of the required computing power in suitable cases.

Accuracy of Molecular Model and Force Field

As was discussed in Section 3.1 .I, there is no best molecular model or force field for all possible applications. The reliability of a particular force field will depend on the type of system and physical quantity it is applied to. Improvement of the quality and extension of the range of applicability is a continual concern in the area of computer simulation. Yet, the improvement of a force field is not a simple exercise for the following reasons.

•   When applying complicated force fields like (I), the exact relationship between a.force,fieldparameter and a molecular property is often not known. For example, how will the free enthalpy of solvation of methanol in water depend on the geometry and nonbonded interaction parameters?

•   Force ,field parameters may be correlated. For example, the actual barrier for a torsional rotation will depend on the combined effect of the dihedral angle potential energy term in (I) and the third neighbor nonbonded interaction between the first and the last atom defining the torsional angle.

- There may be conflicting requirements jor improvement. For example, the van der Waals radii of third neighbor atoms must not be too large in order that the energy of a gauche conformation is not too high in the case of a hydrocarbon chain. However, to obtain the correct density for a hydrocarbon liquid the van der Waals radius must be chosen larger. This conflict may be solved by a separate treatment of third-neighbor and other van der Waals interactions.

- There is some conflict between the wish to stick to conceptual simplicity of a force field on the one hand, and the wish to extend its range of applicability and accuracy by allowing the force field to become more complex.

- Approximations inherent in certain force fields may block substantial improvement by changing of the parameters. For example, when applying a cut-off radius to long-range Coulomb forces, it is impossible to obtain a proper representation of electrostatic effects.

The inclusion of polarizability will be essential for an accurate modeling of processes like the binding of charged ligands. Polarizability only allows for a local displacement of charge, not for a transfer of charge. This can be achieved by combining static ab initio or semi-empirical quantum methods to compute charge densities with classical MD simulation in an adiabatic way: the quantum charge density is converted into atomic charges which are used in a number of classical simulation steps, and the new atomic positions are subsequently used to generate a new quantum charge distribution, and so on.

1911 This type of treatment is only an improvement over conventional simulation when (i) the quantum Hamiltonian includes the long-range electrostatic

field, and (ii) the classical Hamiltonian accounts in a consistent way for the changing charge distribution. In view of the limited accuracy of currently available force fields it may be wise in practical applications to check whether the results obtained are force field dependent.

Non-Equilibrium Molecular Dynamics

In this paper we have considered the computer simulation of systems in equilibrium. It is also possible to change the equations of motion and boundary conditions such that the system is kept in a non-equilibrium state. In such a non-equilibrium molecular dynamics (NEMD) simulation, a nonequilibrium ensemble is sampled. NEMD simulation is an efficient technique to obtain transport coefficients, like viscosity, thermal conductivity, and mobility of molecular systems. An example is the study of the molecular viscosity of n-alkanes as a function of the intramolecular interaction function. For an introduction to NEMD methods and their applications see Refs. It is a rapidly developing field, which is still almost unexplored, with prospects in the area of practical rheology.

Development of Computing Power

The growth of the field of computer simulation of fluid-like systems has been made possible by the steady and rapid increase of computing power over the last couple of decades. Figure 2 suggests an increase of an order of magnitude every 5-7 years. This trend will continue in the near future, since the present growth of computing power is based on the introduction of massive parallelism. The possibilities of parallel computation can be exploited rather easily in MD simulations, since the most time-consuming part is the force calculation, which can be carried out in parallel for all atoms in the system. Also the integration step can easily be performed in parallel. The required computing time for a MD simulation depends linearly on the simulation period and the number of particles (in case of forces of finite range). Therefore this type of calculation can optimally benefit from the continuous growth of computing power. In contrast, quantum mechanical calculations extend their range much more slowly. due to the fact that the required computing power is approximately proportional to N_e^3 in the case of semiempirical or local density functional calculations, to N_e^4 in the case of Hartree-Fock calculations, or even to N_e^5 in the case of configurational interaction calculations. Here, N_e is the number of electronic degrees of freedom. In classical simulations the required computing power is proportional to N_a, or in the case of inclusion of long-range forces or polarizability to at most N_a^2, where N_a denotes the number of atomic degrees of freedom.

SUMMARY AND OUTLOOK

Dynamic computer simulation is just a branch of computational chemistry and physics in which a mathematical model of the real world is formulated and its consequences for the various physical or chemical quantities are evaluated

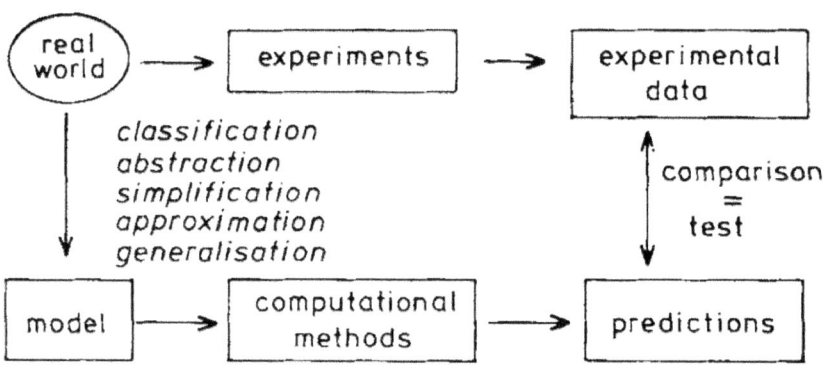

Figure 17: Computational physics and chemistry involve formulation and testing of a (mathematical) model of the real world.

by numerical methods. This is illustrated in Figure 17. In this paper much attention has been devoted to methodology, because the core of any model is the approximations, assumptions and simplifications that are used in its formulation. Only an understanding of the basics of a particular model may lead to sensible application or improvement of its performance. Secondly, attention has been focussed on comparison of model predictions with experimental data, which may reveal flaws in the model (or experiment). Due to the complexity of the systems of chemical interest, theoretical methods only became of practical interest with the advent and development of computers. With the continuous progress of methodology and computing power, computational methods, especially computer simulation methods, will find wider and wider application in the different areas of chemistry.

In Table 7 we have listed our thoughts about the possible development of various aspects of (molecular dynamics)

computer simulation methodology and application in chemistry. From the contents of this paper it will be clear that there are still a number of very difficult problems, such as protein folding or crystallization, which are well beyond the reach of simulation methods owing to the size of the configurational space involved, the time scale of the process, and the small free energy differences between folded and unfolded state or between crystalline and liquid state.

Table 7: Development of various aspects of molecular dynamics computer simulation methods in chemistry.

Aspect	Past (1980)	Present (1990)	Future (2000)
accuracy			
– atomic positions	≈ 3 Å	≈ 1 Å	≈ 0.5 Å
– free energy	–	$\approx 4\, k_B T$	$\approx 2\, k_B T$
force field	united atoms	all atoms	polarizability
environment	vacuo	solvents	membranes
time scale	≈ 10 ps	≈ 100 ps	$\gtrsim 1$ ns
system size	≈ 1000 atoms	$\approx 10\,000$ atoms	$\approx 100\,000$ atoms
quantum (MD) degrees of freedom	–	simple models	enzyme reactions
non-equilibrium simulations	atomic fluids	simple polymers	rheology of molecular mixtures

We would like to thank all our co-workers and colleagues who contributed to the work described here. This work was supported by the Netherlands Foundation for Chemical Research (SON) with financial aid from the Netherlands Organisation for Scientific Research (NWO). The GROMOS (GROningen Molecular Simulation) program library, which can be used for applying the methods discussed here, is available at nominal cost from the authors.

REFERENCES

1.   R W Hockney, J W Eastwood: Computer Sinzulation usmg Particles, McGraw-Hill, New York 1981

2.   J Hermans (Ed.): Molecular Dynamics and Protein Structure Polycrystal Book Service, Western Springs, IL, USA 1985

3.   G Ciccotti, W G Hoover (Eds.): Molecular-Dynamics Simulation of Statistical-Mechanical Systems (Proc lnt School Phys "Enrico Fermi", Cowre 97), North-Holland, Amsterdam 1986

4.   D L Bevendge, W L Jorgensen (Eds.): Compuier Simulation of Chemical and Biomolecular Sysrems (Ann N Y Acad Sci 482 (1986))

5.   J A McCammon S C Harvev: ~, Dvnamics of Proteins and Nucleic Acids Cambridge University Press, London 1987

6.   M P Allen, D J Tildesley: Computer Simulation ofliquids, Clarendon, Oxford 1987

7.   G Ciccotti, D Frenkel, I R McDonald (Eds.): Simulation of Liquids and Solids, North-Holland, Amsterdam 1987

8.   W F van Gunsteren, P K Weiner (Eds.): Computer Simulation of Biomolecular Systems, Escom, Leiden 1989

9.   C R A Catlow, S C Parker, M P Allen (Eds.): Computer Modelling of Fluid Polymers and Solids (NATO ASi Ser C293 (1990))

10.   M.-L Saboungi, A Rahman, J W Halley, M Blander, J Chem Phys 88 (1988) 5818

11.   W L Jorgensen, J Tirddo-Rives, J Am Chem Sac It0 (1988) 1657

12.   W F van Gunsteren, H J C Berendsen: Groningen Molecular Simulution (GROMOS) Library Manual, Biomos, Groningen 1987

13.   N L Allinger, J Am Chem Soc 99 (1977) 8127;

14.   U Burkert, N L: Allinger: Moleculur Mechanics, American Chemical Society, Washington D.C 1982

15.   M Levitt, A Warshel, Nature (London) 253 (1975) 694

16.   B J Berne in: J U Brackbill, B I Cohen (Eds.): Multiple Time Scales, Academic, New York 1985, p 419

17.   J A McCammon, M Karplus, Proc NuII Acad Sci USA 76 (1979) 3585, Biopolymers 19 (1980) 1375

18.   B J Alder, T E Wainright, J Chem Phys 27 (1957) 1208; rbid 31 (1959) 459 A Rahman, Phys Rev 136 (1964) A405

19.   A Rahman, E H Stillinger, J Chem Phys 55 (1971) 3336

20.   L V Woodcock, Chem Phys Lett 10 (1971) 257

21.   J P Ryckaert, A Bellemans, Chem Phys Lett 30 (1975) 123

22.   J A McCammon, B R Gelin, M Karplus, Nature (London) 267(1977) 585

23.   P van der Ploeg, H J C Berendsen, J Chem Phys 76 (1982) 3271

24.   W F van Gunsteren, H J C Berendsen, J Hermans, W G J Hol, J P M Postma, Proc NatL Acad Sci USA 80 (1983) 4315

25.   W F van Gunsteren H J C Berendsen R G Geurtsen H R J Zwinderman, Ann N Y Acad Sci 482 (1986) 287 104

26.   J de Vheg, H J C Berendsen, W F van Gunsteren, Proteins 6 (1989)

27.   G Jacucci, A Rahman, J Chem Phys 69 (1978) 4117

28.   W F van Gunsteren in C Troyanowsky (Ed.): Modelling qf Molecular

29.   J E H Koehler, W Saenger, W F van Gunsteren, Eur Bioph.vs J 15

30.   J Lautz, H Kessler, R Kaptein, W F van Gunsteren J Compur Aided

31.   H J C Berendsen, W F van Gunsteren, H R J Zwinderman, R G

32. A P Heiner, R R Bolt, H J C Berendsen, W F van Gunsteren un-

33. J E H Koehler, W Saenger, W F van Gunsteren, J Biomol S/rurt 1341

34. J E H Koehler, W Saenger, W F van Gunsteren, Eur Brophys J 16

35. R R Ernst, G Bodenhausen, A Wokaun: Principles of Nuclear Mag-

36. J Seelig, Q Rev Biophys 10 (1979) 353

37. E Egberts, H J C Berendsen, J Chem Phy.s 89 (1988) 3718

38. H J C Berendsen, J P M Postma W F van Gunsteren, J Hermans in B Pullman (Ed.): intermolecular Forces, Reidel, Dordrecht 1981 p 331 Structures and Properties, Elsevier, Amsterdam 1990 (1987) 197; ibid 15 (1987) 211 Mol Des 1 (1987) 219 Geurtsen, Ann N Y Acad Sci 482 (1986) 269 published Dyn 6 (1988) 181 (1988) 153 nelic Resonance in One and Two Diinensions Clarendon, Oxford 1987

39. H J C Berendsen, J R Grigera, T P Straatsma, J Phys Chem 91 (1987) 6269

40. W F van Gunsteren, H 3 C Berendsen in J Hermans (Ed.): Molecuiar Dynamics and Protein Structure, Polycrystal Book Service, Western Spnngs, IL, USA 1985, p 5

41. T P Straatsma, H J C Berendsen, J Chem Phys 89 (1988) 5876

42. P Dauber-Osguthorpe V A Roberts, D J Osguthorpe, J Wolff, M

43. F A Mommany, R F McGuire, A W Burgess, H A Scheraga, J

44. M Levitt, J Mol BiOl 168 (1983) 595

45. Z I Hodes, G Nemethy H A Scheraga, BipoJymers 18 (1979) 1565

46. B R Brooks, R E Bruccoleri, B D Olafson, D J States, S Swami-

47. S J Weiner, P A Kollman, D T Nguyen, D A Case, J Comput Chem

48. A T Hagler, E Huler, S Lifson, J Am Chem SOC 96 (1974) 5319

49. J Hermans, H J C Berendsen, W F van Gunsteren, J P M Postma

50. E Platt, B Robson, J Theor Biol 96 (1982) 381

51. W L Jorgensen, J Chandrasekhar, J D Madura, R W Impey, M L

52. D W Rebertus, B J Berne, D Chandler, J Chem Phys 70 (1979) 3395

53. S Lifson, A T Hagler, P Dauber, J Am Chem SOC 101 (1979) 5111

54. G Nemethy, M S Pottle, H A Scheraga, .I Phyr Chem 87 (1983) 1883

55. L G Dunfield, A W Burgess, H A Scheraga, J Phys Chem 82 (1978) 2609

56. K Heinzinger in C R A Catlow, S C Parker, M P Allen (Eds.): Computer Modelling of Fluid Polymers and Solids (NATO AS1 Ser C 293 (1990)) p 357 Genest, A T Hagler, Proteins 4 (1988) 31 Phys Chem 79 (1975)

2361 nathan, M Karplus, J Cornput Chem 4 (1983) 187 7 (1986) 230 Biopolymers 23 (1984) 1513 Klein, J Chem Phys 79 (1983) 962

57. A Paskin, A Rahman Phys Rev Letr 16 (1966) 300

58. S N Ha, A Giammona, M Field, J W Brady, Carhohydr Res f80

59. D van Belle, I Couplet, M Prevost, S Wodak, J Mol Biol 198 (1987)

60. F J Vesely J Comput Phys 24 (1977) 361

61. B J Alder, E L Pollock, Annu Rev Phys Ckem 32 (1981) 311

62. P Barnes, J L Finney, J P Nicholas, J E Quinn, Nature (London) 282

63. A Warshel, H Levitt, J Mol Biol 103 (1976) 227

64. S T Russell, A Warshel, J Mol Biol I85 (1985) 389

65. W G J Hol, P T van Duynen, H J C Berendsen, Nature (London) 273 (1978) 443

66. H J C Berendsen in J Hermans (Ed.): Molecular Dynamics and Protein Slructure, Polycrystal Book Service, Western Springs, IL, USA 1985 18

67. A Warshel, S T Russel, Q Rer Biophys 17 (1984) 283

68. B R Brooks in K F Jensen, D G Truhlar (Eds.): Supercomputer Research in Chemistry and Chemicai Engineering (ACS Symp Ser 353 (1987) 123)

69. J Aqvist, W F van Gunsteren, M Leijonmarck, 0 Tapia, J Mol Biol 183 (1985) 461

70. A J C Ladd, Mol Phys 33 (1977) 1039; ibid 36 (1978) 463

71. H L Friedman, Mol Phys 29 (1975) 1533

72. J Warwicker, H C Watson, J Mol Biol 157 (1982) 671

73. J P van Eerden in W Olson (Ed.): Nucleic Acid Conformarion and Dynamics (Rep NATOjCECAM Workshop), CECAM, Orsay 1983 p 61

74. P B Shaw, Phys Rev A32 (1985) 2476

75. R J Zauhar, R S Morgan, J Mol Biol 186 (1985) 815; J Comput

76. P Ewald, Ann Phys (Leipzrg) 64 (1921) 253

77. S W de Leeuw J W Perram, E R Smith, Proc R Sac London A373

78. D M Heyes, 1 Chem Phys 74 (1981) 1924

79. K B Wiberg, R H Boyd, J Am Chem Soc 94 (1972) 8426

80. M Lipton, W C Still, J Comput Chem 9 (1988) 343

81. J Moult, M N G James, Proteins I (1986) 146

82. K D Gibson, H A Scheraga, J Comput Chem 8 (1987) 826

83. S H Northrup, J A McCammon, Biopolymers 19 (1980) 1001

84. G M Crippen in D Bawden (Ed.): Chemometric Research Studres Se-

85. T Havel, 1 D Kuntz, G M Crippen, BuN Math Biol 45 (1983) 665

86. T Havel, K Wiithrich, Bull Math Bfol 46 (1984) 673

87. R P Sheridan, R Nilakantdn J S Dixon, R Venkataraghavan, J Med

88. A DiNola H J C Berendsen, 0 Edholm, Macromolecules 17 (1984)

89. M Berkowitz, J A McCammon, Chem Phys Leu 90 (1982) 215

90. W F van Gunsteren, H J C Berendsen, J A C Rullmann, Mol Phw

91. Shi Yun-yu, Wang Lu, W F van Gunsteren, Mol Simularion I (1988)

92. T Akesson, B Jonsson, Mol Phys 54 (1985) 369

93. S Kirkpatrick C D Gelatti Jr., M P Vecchi, Science (Washington

94. R A Donnelly, J W Rogers, Jr., Int J Quantum Chem.: Quantum

95. 2 Li, H A Scheraga, Proc Nutl Acad Sci USA 84 (1987) 6611

96. W Braun, N Go, J Mol Biol 186 (1985) 611 (971

97. W F van Gunsteren, M Karplus, Biochemistry 21 (1982) 2259

98. G L Seibel, U C Singh, P A Kollman, Proc Natl Acad Sci USA 82

99. K Remerie, W F van Gunsteren, J B F N Engberts, Chem Phys 101 D.C.j
     220 (1983) 671 Chem Svmp 22 (1988) 507 (1985) 6537 (1986) 27

100. M J Mandel!, J Slat Phys 15 (1976) 299

101. L R Pratt, S W Haan, J Chem Phys 74 (1981) 1864, 1873

102. S Gupta, J M Haile, Chem Phys 83 (1984) 481

103. D J Adams, Chem Phys Lett 62 (1979) 329

104. J 9 Gibson, A N Goland, M Milgram, G H Vineyard, Phys Rev

105. C L Brooks 111, M Karplus, J Chem Phys 79 (1983) 6312

106. A T Briinger, C L Brooks 111, M Karplus, Chem Phys Lett 105

107. C L Brooks 111, A Briinger, M Karplus, Biopolymers 24 (1985) 843

108. A T Brunger C L Brooks 111, M Karplus, Proc Narl Acad Sci USA

109. W G Hoover, D J Evans, R B Hickman, A J C Ladd, W T Ashurst,

110. D J Evans, J Chem Phys 78 (1983) 3297

111. J M Haile, S Gupta, J Chem Phys 79 (1983) 3067 I

112. S Nose, Mol Phys 52 (1984) 255

113. H J C Berendsen, J P M Postma, W F van Gunsteren, A DiNola, J R
     Haak, J Chem Phys 81 (1984) 3684

114. H C Andersen, J Chem Phys 72 (1980) 2384

115. D M Heyes, Chem Phys 82 (1983) 285

116. T Schneider E Stoll Phys Rev 817 (1978) 1302; ibrd B18 (1978)

117. D J Evans, G P Morriss, Chem Phys 77(1983) 63; Comput Phys Rep

118. J M Haile, H W Graben, J Chem Phys 73 (1980) 2421

119. M Parrinello, A Rahman, Phys Rev Lett 45 (1980) 1196; J Appl Phys

120. M Parrinello, A Rahman, P Vashishta, Phys Rev Lett 50 (1983)

121. S Nose, M L Klein, Mol Phys 50 (1983) 1055

122. W H Press, B P Flannery, S A Teukolsky, W T Vetterling: Numerical Recipes Cambridge University Press, Cambridge, England 1986

123. C W Gear, Numerical Initial Value Problems in Ordinary Drfferenrial Equations, Prentice-Hall, Englewood Cliffs, NJ, USA 1971

124. L Verlet, Phys Rev 159 (1967) 98

125. D Beeman, J Comput Phys 20 (1976) 130

126. W F van Gunsteren, H J C Berendsen, Mol Phys 34 (1977) 1311

127. H J C Berendsen, W F van Gunsteren in G Ciccotti, W G Hoover (Eds) Mofeculur-Dvnamics Simulation of Statistical-Merhanics Sysfems, North-Holland, Amsterdam 1986, p 43

128. W F van Gunsteren H J C Berendsen, Mol Simulation 1 (1988) 173

129. G Ciccotti, M Ferrario, J-P Rijckaert, Mol Phys 46 (1982) 875

130. F J Vesely Mol Phvs 53 (1984) 505

131. H J C Berendsen W F van Gunsteren in J W Perram (Ed.): The Phwcs of Superionic Conductors and Electrode Materials (NATO ASI Sm B92 (1983)) 221 120 (1960) 1229 (1984) 495 82 (1985) 8458 9 Mordn, Pliys Rev A22 (1980) 1690 6468 1 (1984) 297 52 (1981) 7182; J Chem Phys 76 (1982) 2662 1073

132. W F van Gunsteren, Mol Phys 40 (1980) 1015

133. J Barojas, D Levesque, B Quentrec, Phys Rev A 7 (1973) 1092

134. P S Y Cheung, Chem Phys Lett 40 (1976) 19

135. D J Evans, S Murad, Mol Phys 34 (1977) 327

136. M Fixman, J Chem Phys 69 (1978) 1527]

137. J Wittenburg: Dynamrcs of Systems of Rigid Bodies Teubner, Stuttgart

138. J P Ryckaert G Ciccotti H J C Berendsen, J Comput Phys 23

139. W F van Gunsteren, M Karplus, Macromolecules 15 (1982) 1528

140. G Ciccotti, M Ferrario, J P Ryckaert, Mol Phy.s 47 (1982) 1253

141. W F van Gunsteren in W F van Gunsteren, P K Weiner (Eds.) Com-

142. W B Street, D J Tildesley, G Saville, Mo/ Phys 35 (1978) 639

143. 0 Teleman, B Jonsson J Comput Chem 7 (1986) 58

144. W F van Gunsteren, H J C Berendsen, F Colonna, D Perahia, J P 1977 (1977) 327 puter Simulution of Biomoleculur Systems, Escom, Leiden 1989, p 27 Hollenberg, D Lellouch, J Cornput Chem 5 (1984) 272

145. R Boelens, R M Scheek, J H van Boom, R Kaptein J Mol Biol 193

146. B W Matthews, Nature (London) 335 (1988) 294

147. M H Caruthers, Arc Chem Res 13 (1980) 155

148. J H Miller, J Mol Biol 180 (1984) 205

149. H Schindler, J Seelig, J Chem Phys 59 (1973) 1841

150. R Kaptein, R Boelens, R M Scheek, W F van Gunsteren, Eiochemistry 27 (1988) 5389

151. W F van Gunsteren, R Kaptein, E R P Zuiderweg in W K Olson (Ed.) Nucleic Acid Conformation and Dynumics (Rep NATOICECAM Workshop) CECAM, Orsay 1984, p 70 (1987) 213

152. G M Clore, A M Gronenborn, Protein Eng I (1987) 275

153. J Lautz, H Kessler, J M Blaney R M Scheek, W F van Gunsteren,

154. M Nllges, G M Clore, A M Gronenborn, FEBS Lerf 239 (1988) 29

155. H Kessler, C Griesinger, J Lautz, A Miiller, W F van Gunsteren, H J C Berendsen, J Am Chem Soc 110 (1988) 3393

156. A E Torda, R M Scheek, W F van Gunsteren, Chem Phm Lett 157 (1989) 289

157. A T Briinger, J Kuriyan, M Karplus, Science (Washingron D.C.j 235 (1987) 458

158. P Gros, M Fujinaga, A Mattevi, F M D Vellieux, W F van Gunsteren, W G J Hol in J Goodfellow, K Henrick, R Hubbard (Eds.): Molecular Simulation and Protein Crystullography (Proc Joint CCP4I CCPS Study Weekend) SERC, Daresbury 1989, p 1

159. P Gros, M Fujinaga, B W Dijkstra, K H Kalk, W G J Hol, Acta Crjvtallogr 845 (1989) 488

160. R R Bott, WorldBiotech Rep 2 (1985) 51

161. W F van Gunsteren, H J C Berendsen, J Mol Biol 176 (1984) 559

162. P Kriiger, W Strassburger, A Wollmer, W F van Gunsteren Eur Biophys J 13 (1985) 77

163. J E H Koehler, W Saenger, W E van Gunsteren, J Mol Bid 203 (1988) 241

164. J Lautz, H Kessler, W F van Gunsteren, H P Weber, R M Wenger, Biopolymers (1990), in press

165. D L Beveridge, F M DiCapua, Ann Rev Biophys Biophys Chem 18 (1989) 431

166. D A Matthews, J T Bolin, J M Burridge, D J Filman, K W Volz J Kraut, J Biol Chem 260 (1985) 392

167. D A Matthews, J T Bolin, J M Burridge, D J Filman K W Volz B T Kaufman, C R Beddell, 1 N Champness, D K Stammers, J Kraut, J Biol Chem 260 (1985) 381 Int J Pept Protein Res 33 (1989) 281

168. B Klar, B Hingerty, W Saenger, Acra Crystallogr B36 (1980) 1154

169. Z Otwinowski, R W Schevitz, R G Zhang, C L Lawson, A Joachimiak, R Q Marmorstein 9 F Luisi, P 9 Sigler, Nature (London) 335 (1988) 321

170. V Vlachy D J Haymet, J Chem Phys 84 (1986) 5874

171. J Tallon, Phys Rev Leu 57 (1986) 2427

172. R W Impey, M Sprik, M L Klein, J Chem Phys 83 (1985) 3638

173. F F Abraham, W E Rudge, D J Auerbach, S W Koch, Phys Rev

174. D M Heyes, M Barber, J H R Clarke, Surf Sci I05 (1981) 225

175. V Rossato, G Ciccotti, V Pontikis, Ph.vs Rev 833 (1986) 1860

176. U Landman, W D Luedtke, R N Barnett, C L Cleveland, M W Ribarsky, E Arnold, S Ramesh, H Baumgart, A Martinez, B Khan, Phys Rev Leu 56 (1986) 155 Letf 52 (1984) 445

177. J G Powles, G Rickayzen, Mol Phys 38 (1979) 1875

178. W G Gibson Mol Phys 28 (1974) 793

179. P H Berens, D H J Mackay, G M White, K R Wilson, J Chem

180. B J Berne, G D Harp, Adv Chem Phys 17 (1970) 63

181. J A Barker, J Chem Phys 70 (1979) 2914

182. D Chandler, P G Wolynes, J Chem Phys 74 (1981) 4078

183. B de Raedt, M Sprik, M Kleln, J Chem Phys 80 (1984) 5719

184. C Zheng, J A McCammon P G Wolynes, Pror Null Acad Sci USA

185. E J Heller, J Chem Phys 62 (1975) 1544 and 64 (1976) 63

186. N Corbin, K Singer, Mol Phys 46 (1982) 671

187. K Singer, W Smith, Mol Phys 57 (1986) 761

188. R Carr, M Parrinello, Phys Rev Lett 55 (1985) 2471

189. A Selloni P Carnevali, R Carr, M Parrinello, Phys Rev Lett 59

190. C H Bennett, J Comput Phys 19 (1975) 297

191. P A Bash, M J Field, M Karplus, J Am Chem Soc f09 (1987) 8092

192. J H R Clarke, D Brown, J Chem Phys 86 (1987) 1542.

CITATION

CHAPTER 1

Alexander Pertsin, Dmitry Platonov, and Michael Grunze, Computer simulation of short-range repulsion between supported phospholipid membranes, DOI: 10.1116/1.2190699

CHAPTER 2

Michael R. Twiss , Olivier Errécalde , Claude Fortin , Peter G. C. Campbell , Catherine Jumarie , Francine Denizeau , Edward Berkelaar , Beverley Hale , Ken van Rees, Coupling the use of computer chemical speciation models and culture techniques in laboratory investigations of trace metal toxicity, Vol. 13, Iss. 1, 2001

CHAPTER 3

P. A. Netz; F. Starr; M. C. Barbosa; H. Eugene Stanley, Computer simulation of dynamical anomalies in stretched water, http://dx.doi.org/10.1590/S0103-97332004000100004

CHAPTER 4

Zhang Y, Thomas GL, Swat M, Shirinifard A, Glazier JA (2011) Computer Simulations of Cell Sorting Due to Differential Adhesion. PLoS ONE 6(10): e24999. doi:10.1371/journal.pone.0024999

CHAPTER 6

Yong-Lei Wang, Fredrik Hedman, Massimiliano Porcu, Francesca Mocci, Aatto Laaksonen, Non-Uniform FFT and Its Applications in Particle Simulations, http://dx.doi.org/10.4236/am.2014.53051

CHAPTER 7

Arkady Ilyin (2011). Computer Simulation of Radiation Defects in Graphene and Relative Structures, Graphene Simulation, Prof. Jian Gong (Ed.), ISBN: 978-953-307-556-3, InTech, DOI: 10.5772/20338.

CHAPTER 8

Wilfred E van Gunsteren and Herman J. C. Berendsen, Computer Simulation of Molecular Dynamics:Methodology, Applications, and Perspectives in Chemistry, doi/10.1002/anie.199009921/

INDEX